VICTORY GARDENS
for Bees

A DIY GUIDE TO SAVING THE BEES

LORI WEIDENHAMMER

VICTORY GARDENS
for Bees

A DIY GUIDE TO SAVING THE BEES

Douglas & McIntyre

1 2 3 4 5 — 20 19 18 17 16

Douglas and McIntyre (2013) Ltd.
P.O. Box 219, Madeira Park, BC, VON 2H0
www.douglas-mcintyre.com

PHOTOS BY Lori Weidenhammer except where otherwise noted
EDITED BY Carol Pope and Nicola Goshulak
INDEXED BY Kyla Shauer
TEXT AND COVER DESIGN BY Mauve Pagé
PRINTED AND BOUND IN CANADA
Printed on paper certified by Forest Stewardship Council

Douglas and McIntyre (2013) Ltd. acknowledges the support of the
Canada Council for the Arts, which last year invested $157 million to
bring the arts to Canadians throughout the country. We also gratefully
acknowledge financial support from the Government of Canada through
the Canada Book Fund and from the Province of British Columbia
through the BC Arts Council and the Book Publishing Tax Credit.

Cataloguing data available from Library and Archives Canada
ISBN 978-1-77162-053-6 (paper)
ISBN 978-1-77162-054-3 (ebook)

Previous spread: A red-belted bumblebee (*Bombus
rufocinctus*) forages in oregano. **Following page:** Sunflower
blossoms are like buffets for bees, with plenty of room for
dining cheek to cheek. **Page vi–vii spread:** A mixed bumblebee
(*Bombus mixtus*) forages in a blackberry blossom.

This book is dedicated to the gardeners, farmers and ecologists of the future. May their lives be filled with the murmuring of innumerable bees. And to Mom and Dad, who gave my sister and me a childhood abundant with prairie wildflowers, bees and plenty of time for reverie.

Acknowledgements

"Where flowers bloom, so does hope." —LADY BIRD JOHNSON

I would like to thank my editor, Carol Pope, for taking the leap of faith to undertake this project. The credit for this book should be shared between Carol and me, with her acting as a patient and generous mentor and collaborator. Thanks to my husband, Peter, and our son for allowing the house to become chaotic with growing piles of reference books and draft pages, for tolerating the dishes piled in the sink and negotiating the bags of seeds stuffed into every nook and cranny of our abode. Thanks to the friends who suggested I should write a book, and celebrated with me when the opportunity came. Thanks to all the gardeners, photographers, artists, scientists and bee huggers who shared their wisdom and enthusiasm with me and allowed us to print their work. Thanks to the crew at Shaktea who poured pots of tea and murmured encouragement as I toiled week after week in their cozy shop.

Thanks to Jasna Guy for lending me her fabulous camera, helping with some tough proofreading and taking me on bee safaris to keep me sane and refreshed. Thanks to my mentor Brian Campbell for his sweet wisdom, deep knowledge, humour and support. Thanks to my neighbours Jean Kindratsky and Catherine Shapiro for gardening wisdom and formative adventures in backyard beekeeping. Thanks to Erin Udal for helping me with bee identification and supporting my efforts all along the way. Thanks to John Ascher, bee identification expert at BugGuide.net. Thanks to Mark Wonneck for pulling over in his truck in the middle of Alberta to chat with me about bees. Thanks to Len and Mark for helping my writing hand heal after the "climbing" accident. Thanks to all the folks whose quotes and recipes I solicited but whose contribution did not make the final cut—there were so many brilliant ideas I wanted to include, but alas, there was not enough space.

A special thanks to the team at Douglas and McIntyre, especially Nicola, as assistant editor and all the other hats she wears. Any mistakes are my own. Check in on my blog (www.beespeakersaijiki.blogspot.ca) for any corrections and updates.

Opposite: Besides providing food for bees, Joe-Pye weed has hollow stems that provide excellent homes for stem-nesting bees.

Preface

*E*nvision a world without bees. A world where 80 percent of flowers fade away. Where three-quarters of food crops vaporize. Where not a single bee thrums, apple trees stand bare, almonds and avocados are extinct, cherries and blueberries only dim recollections from days gone by.

It is not just chance that bees pollinate flowers: bee and bloom synergistically develop at the precise moments necessary for their shared survival. But now global warming is inflicting havoc on time-tested weather patterns, habitat loss is stripping bees of precious homeland, hives are in calamity due to Colony Collapse Disorder and bees far and wide are plagued by a deluge of parasites and deadly pesticides.

More than half a century ago, the threat of widespread famine loomed amongst the many terrors of World Wars I and II. Yet, instead of resigning themselves to scarcity, citizens rolled up their sleeves. In 1943, in North America alone, Victory Gardens in backyards, window boxes, patio planters, schoolyards, parks and government grounds burst forth with nearly 9 million tons of food, much of it coaxed from the ground by people who had never before grown a carrot or cauliflower.

Today, there is need for a similar call to action. Only by planting a surge of Victory Gardens for Bees will we resolve the desperate famine of forage and shelter faced by pollinators. As author Lori Weidenhammer—beespeaker, guerrilla gardener, multimedia artist and educator—teaches us in this essential guide to saving bees, a communal effort to create pollen- and nectar-rich pit stops, plant pathways and bee accommodation is vital if we want to avert a "beepocalypse."

From hive honeybees to bombastic bumblebees to stingless solitary bees, *Victory Gardens for Bees* is the essential DIY directive to saving bees of all stripes. With ten inspiring full-colour planting plans designed specifically for bees, along with extensive photography, easy-reference four-season plant guides and indispensable tips for identifying and helping bees throughout, this book is what we all need to help them survive and thrive.

Whether we grow a lush beekeeper's garden, an edible bed of umbels and orbs, healing herbs for pollinators, indigenous plants, a heritage hedgerow or fragrant balcony containers, or even simply take note of what weeds to leave for bees, we can create the change bees need if they are to endure.

As we each pick up a shovel and seeds to create Victory Gardens for Bees, we literally hold the fruitfulness of earth in our hands. We *can* be the difference between whether these inspiring and beautiful creatures flounder or flourish. And flourish they must, for the future of this planet.

—Carol Pope
    Associate Editor

Opposite: A mass of blazing star will attract scores of bumblebees to your garden. Plant it instead of invasive thugs such as purple loosestrife (*Lythrum salicaria*).

Introduction

*A*s the planet groans under the weight of human population growth, we should be gardening for our lives. We need to undo the damage we have done to make the world inhospitable for the very bees that make our planet liveable—and it is through our gardens that each of us can make a difference.

During World War II, the Allied forces created propaganda posters to convince people on the home front to grow Victory Gardens. A Canadian World War II poster with an illustration of marching vegetables declared, "Help Canada and have fun too! Grow these 'fighting foods' at home." In the United States, people could put a patriotic decal in the window that proudly stated, "We have a Victory Garden." A British poster suggested pointedly, "Every available piece of land should be productive. 'Scuse the dig, but what about your lawn?" In the end, the Dig for Victory campaigns were so overwhelmingly successful that even the Allied War Offices were taken by surprise.

When it comes to supporting bees, it's time we did the math. The biomass of bees on our planet needs a critical amount of space to raise their brood, and enough pollen and nectar to feed this offspring and fuel their entire adult cycle. If we want to protect our pollinators and the very survival of life on our planet, we need to grow Victory Gardens for Bees with a level of dedication similar to the Victory Gardens of our past. With over one-third of our food crops and 80 percent of flowering plants pollinated by bees, it is essential we join efforts to ensure their protection and survival.

—*Lori Weidenhammer*

The New Victory Gardens

Joe Wirtheim is a designer who has reinvented the original Victory Garden posters as compelling images for the contemporary movements that appreciate the value of growing food close to home. Wirtheim channels the positive "can-do" voice of the posters originally used to inspire folks on the home front to take action and "Dig for Victory." Only this time around, people are taking up their shovels to create what the artist calls "healthier and happier communities."

JOE WIRTHEIM PHOTO

Opposite: A long-horned bee (*Melissodes*) forages on a brown-eyed Susan blossom.

Bee Afraid

CHAPTER ONE

*I*magine a future where the only bees that children see are pinned specimens in dusty museums, where the sound of bees collecting pollen and sipping nectar fades from human memory, and wind-blown wildflowers exist only in photographs and films. Imagine dinner plates and supermarkets drained of the range of colour and flavour that comes from bee-pollinated herbs, nuts, fruits and vegetables.

Our honeybees are disappearing in numbers difficult to comprehend. Over the past decade, beekeepers in North America and Europe have been losing an average of 30 percent of their hive populations each year, with the adult bees suddenly leaving the hive, never to return. Scientists name the mystery "Colony Collapse Disorder." *TIME* magazine declared it a "beepocalypse."[1] Dr. Mark Winston, one of Canada's pre-eminent honeybee scientists, calls the inside of a typical beehive a "toxic soup" containing over 100 pesticides.[2] And, according to an article by Gwen Pearson in *WIRED* magazine, the problem is bigger than that of the honeybees. Over 50 percent of native bee species in the agriculturally intensive "bee killing fields" of Midwestern America have been lost in the last century.

The Xerces Society's Red List of threatened and endangered bees lists 57 species from Alaska to Hawaii. The western bumblebee, formerly one of the most common wild bees in western North America, is now faced with probable extinction, and three other species of North American bumblebees are gone forever.[3]

To paraphrase the words of a recent Environmental Justice Foundation poster, if the bees disappear, they're taking us with them. Not only food systems for humans would collapse if we lost the bees, but also sustenance for wildlife. No bees, no berries. No berries, no bears. And so on. Bees create the seeds that ensure the survival of many plant communities and the continuation of life on this planet. No bees, no seeds.

No seeds, no future.

Opposite: Italian bugloss (*Anchusa azurea*), which is in the borage family, has striking blue flowers that bumblebees drool over.

The Honeybee Crisis: What Is Colony Collapse Disorder?

In late 2006, the term "Colony Collapse Disorder" (CCD) was coined to describe the unprecedented catastrophic loss of honeybee colonies. Many scientists agree that CCD is the result of a negative synergy of multiple problems: varroa mites, malnutrition from monoculture crops, transportation stress, loss of genetic diversity, pesticides and fungicides, loss of biodiverse habitat, climate change, and compromised immunity. As hive after hive collapses, it is also becoming clear how vital a part our native bees play in pollinating the food we eat.

A Gardening Victory Worth Repeating

During World War II, previously unproductive land was transformed into fruitful community gardens due to the work of ordinary citizens. Margaret RainbowWeb, a woman who grew up in England during World War II, recalls that her grandfather's Victory Garden fed six families, plus their chickens and rabbits. Imagine how many bees it supported as the herbs, fruit trees, berry bushes and self-seeding vegetables blossomed. Imagine the difference that could be made today by Victory Gardens for Bees springing up across North America to feed bees.

NO BEES, NO SEEDS

Eighty percent of flowering plants are pollinated by bees. And not just European honeybees, which are only one of the world's estimated 20,000 species of bees. A commonly quoted statistic says that one out of every three mouthfuls of food you eat is pollinated by bees, but depending on your diet, this could be much higher, since bees pollinate 75 percent of fruits, nuts and vegetables grown in North America. Bees also pollinate the alfalfa that provides food for dairy and beef cattle. Many of the most nutritious food plants—those that provide essential nutrients for human health—have evolved in a close synergistic relationship with bees. And the flowers specifically designed to suit the needs of the bees, providing nectar and pollen as food to nourish them, depend on bees for pollination.

CLIMATE CHANGE AND POLLINATORS: A TIME BOMB

"Don't you find it wonderful that it is completely by accident that bees pollinate flowers?" a visitor to my garden once mused . . .

The truth is that the very special mutual relationship between bees and plants is the result of millennia of evolutionary adaptations. Squash bees, for example,

> "The world as we know it would not exist if there were no bees to pollinate the earth's 250,000 flowering plants."[4]
>
> —BEATRIZ MOISSET AND STEPHEN BUCHMANN, authors of *Bee Basics: An Introduction to Our Native Bees*

evolved to eat, sleep, mate and procreate in the pump-
kin patches they pollinate. Bumblebees nest in the
forest edges that support the blueberries that provide
food for their brood. In fact, they have a special trick to
release pollen called "buzz pollination," vibrating the
flower to release pollen, making the berries healthy
and plump.

Climate change threatens to unhinge this com-
plexly synchronized relationship between pollinators
and their habitat. Plants need to bloom at the times
that coincide with the bees' foraging cycle, or it is disas-
trous for the bees, particularly those designed for pollinating a narrow group of plants. Global warming may
cause these life cycles to shift and fall out of synch with
each other. This means the missing of a crucial set of

Your garden can become a battleground when bees are
competing for precious nectar and pollen. This photo series
shows a sunflower bee (*Svastra*) karate-kicking a long-horned
bee (*Melissodes*) out of a Mexican sunflower. KATHY KEATLEY
GARVEY PHOTOS

appointments for both—with pollinators going hungry and plants failing to fruit and set seed.

CHOOSING FOODS FROM BEE WISE FARMS

As we lose native pollinators and honeybee health declines, crops and wildflowers are suffering an increasing pollination deficit. This means there is a lack of pollinators to serve our agricultural needs and preserve our ecologically sensitive wilderness. This deficit is getting so serious that it threatens the ecology of the earth's biosphere. In Canada, we have an Ocean Wise program, created to empower businesses and consumers to know when they are buying sustainable fish. It's time we implemented a Bee Wise program[6] to acknowledge which foods have been grown by organic farmers who implement agroecology and set aside a third of their land as bee habitat.

> ## *Bee Mitzvah: Plant a Bee Garden Today*
>
> What do you and I have in common with prime ministers, queens, bankers, butchers, bakers and candlestick makers? We all eat foods pollinated by bees—not just those exquisite amber sisters that give us honey, but all bees, from the tiniest *Perdita minima* to the large, sonorous bumblebees that make our blueberries plump and ripe. Bees unite us all and connect us to a shared responsibility to make our yards, parks, farms and boulevards into bee-supportive environments. Making the world a better place for bees is what I call a "bee mitzvah," a good deed that benefits the *bee*stower.

Opposite, clockwise from top left: A mining bee (*Andrena prunorum*) and soldier beetle sip nectar from the tiny florets of masterwort. A glittering jewel wasp snoozes on a stem (SEAN MCCANN PHOTO). Male bees often have longer antennae than females, and some sport little moustaches or goatees (ANNA HOWELL PHOTO). Bird's eye gilia produces powder-blue pollen loved by bees of all stripes (KATHY KEATLEY GARVEY PHOTO). **Below:** Choose Bee-Wise fruit from growers who create habitat that feeds and shelters native bees.

SEEKING A SOLUTION:
VICTORY GARDENS FOR HOPE

All across the globe, more and more people of all ages want to plant gardens to grow their own food, with the desire to reconnect with the miracle of how bees help plants flourish. Many wish to learn how to grow plants that support bees and provide them with pesticide-free food and safe places to shelter. These oases for bees will offer hope for our ailing pollinators, particularly if there are enough gardens to form a network of vibrant pollinator pathways linking healthy habitat for bees.

Top: Research by Dr. Laurence Packer's lab at York University indicates that bees are at a very high risk of extinction compared to most other organisms. **Bottom:** As children, my sister and I gathered prairie crocus each spring; sadly, these bee-supporting wildflowers are now disappearing.

PLANTING PLAN❋

The Community Bee Garden: A Gathering Place to Celebrate Pollinators

It's in our nature to garden together, and we need more pollinator-friendly spaces to share our love of growing and eating local bee-pollinated foods. In this community gathering place, grandparents and toddlers can harvest mint and anise hyssop to make freshly brewed tea. New gardeners work alongside mentors who can happily share their tips on how to prune and divide the lavender and sage. A community garden unites generations through their common love of tasty food and commitment to a healthy bee-supportive habitat. See opposite page for a sample community bee garden plan.

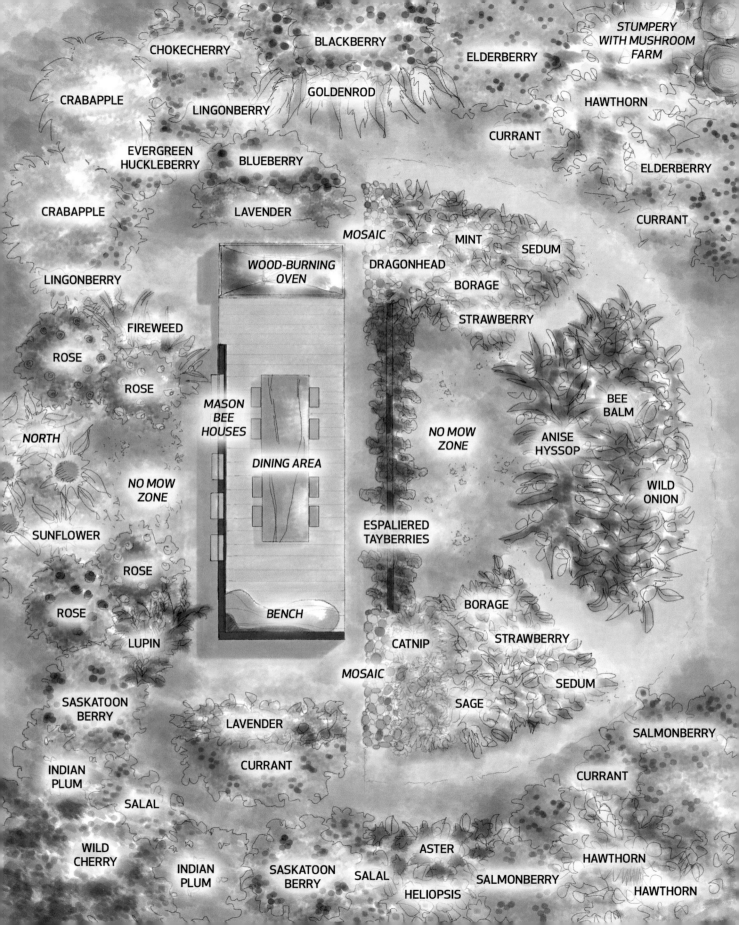

Victory Gardens for Bees

CHAPTER TWO

*J*f you don't know how to begin with your Victory Garden for Bees, start off by considering where you could squeeze in just one pot of lavender, perhaps on a porch or front stoop. As bee habitats become increasingly fragmented, even small links between bee-friendly landscapes make a big difference. Bees that travel long distances to forage, such as honeybees and bumblebees, can get stranded if there are not enough sources of nectar to give them the energy to return home to their nests. Bees also need to travel to mate in order to ensure genetic diversity. From a single pot of French or English lavender to a pollinator system of hedgerows and meadows, we can keep bees criss-crossing country and cityscapes from one food stepping stone to another.

After you have added that lavender plant, plot out where you could plant four more. Or six. Then start thinking bigger. Yes, dig up your lawn if you have one.[7] Toss wildflower seeds onto your boulevard. Add borage and thyme to your raised bed at the community garden and let a few of the garlic plants flower. Give pots of asters and packages of crimson clover seeds to your neighbours. Plant a school garden to teach children about supporting bees and why it matters. Big or small, when you build a bee garden, the bees will come. Every vacant lot, silent lawn or barren balcony has the potential to become a bee-loud glade.

Opposite: The plush centres of elecampane flowers provide luxurious landing and refueling pads for bees.

- **OLD WORLD BEES** are species of bees indigenous to the eastern hemisphere that evolved with Old World plants.
- **NEW WORLD BEES** are species of bees indigenous to the western hemisphere that evolved with New World plants.
- **A MELLIFEROUS PLANT** produces **nectar** that gives bees energy. In the context of beekeeping, this describes a plant that honeybees use to make honey. **Nectar** is the sugar-rich liquid secreted by plants to attract pollinators; differing somewhat from plant to plant in sweetness and viscosity, nectar provides bees with the energy they need to do their life's work.
- **A POLLENIFEROUS PLANT** produces pollen. **Pollen** is the male energy of the plant, half the genetic information of a plant delivered by the flower's stamens. It hitches a ride on the pollinator, which delivers it to the female receptacle of the plant, the pistil (composed of stigma, style and ovary). Bees collect pollen and store it on their bodies to carry back to their nest to feed to their young. Pollen grains vary widely in size and shape and are essential to the development of a bee from egg to larva to adult.
- **A HONEY PLANT** supports the life cycle of honeybees, and often other bee species too, with nectar and/or pollen.
- **A BEE PLANT** provides the food and/or shelter needs of least 1 of the nearly 20,000 species of bees.

A GOOD BEE PLANT— WHAT *DOES* THAT MEAN?

A good bee plant:
- is not invasive and integrates well into a bio-plan[8] that builds biodiversity
- has a rich reward-to-plant-mass ratio, (a high volume of nectar and pollen for the amount of space taken up by the plant)
- blooms continuously over a long period of time, giving it successional value, or replenishes nectar many times over the course of a day
- has a structure that protects pollen and nectar for

A Balcony Garden for Bees

A garden balcony can be an oasis of calm in the bustling city for both you and the bees. Sip your morning latte surrounded by brightly coloured bee plants in jewel-toned ceramic pots. Pick a warm tomato off the vine, along with a few leaves of basil, to make yourself a delicious salad or sandwich. Kick back at dusk with a glass of wine and watch the bumblebees forage in your dwarf sunflower and lavender. If your balcony is cool and shady, grow a variety of mints to muddle in your mojitos and choose native understorey plants like evergreen huckleberry and wintergreen that thrive in the shade. If you are on a hot and sunny side of the building, load your pots with heat-loving herbs such as rosemary and oregano, along with peppers and tomatoes. Climbing runner beans and hanging baskets of nasturtiums help to create a privacy screen and lure bees upward to visit your solitary retreat. See opposite page for a sample plan.

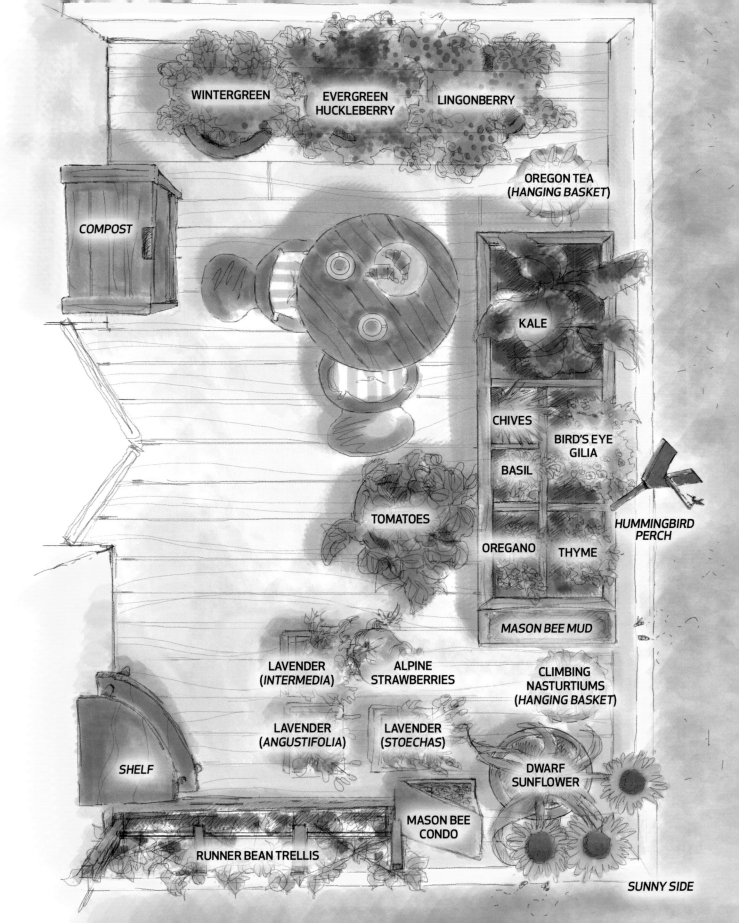

bees in inclement weather, providing consistent rewards

- has a shape that enables cost-efficient foraging within a bee's energy budget
- produces a flower ergonomically and efficiently suited to its pollinators
- has an alluring bee-attracting scent and striking visual design that clearly signals the location of rewards to pollinators
- supports native bees
- has nutritional and medicinal value for bees, and may also have nutritional and medicinal value for humans
- sustains more than one kind of fauna and has a complex ecological function, or supports a rare oligolectic (specialist) pollinator
- is hardy and water-wise and appropriate to the zone where planted
- is free from car-exhaust particulates, pesticides, fungicides, herbicides and miticides
- can supply oils, petals, leaves and resins that serve the needs of bees

THE DIY GUIDE TO GROWING YOUR VICTORY GARDEN FOR BEES

There are 16 essentials when it comes to creating a safe garden space that will provide bountiful benefits for bees:

1. SHUN NEONICS AND OTHER BEE-KILLERS

In 2013, shoppers in a mall parking lot in Wilsonville, Oregon, were horrified to find thousands of bees, mostly bumblebees, dying on the pavement, some still clinging to the linden blossoms they were pollinating when they were sprayed with pesticides.[9] Because they were sticky with aphid sap, the linden trees had been

Neonics: We Need to Ban These Lethal Nerve Toxins

"It's time to ban neonics. These pesticides are nerve toxins."[10] These words from Dr. Elizabeth Elle, professor of biology at Simon Fraser University in British Columbia, are a clear warning to home gardeners and farmers concerned about the health of bees. Recent restrictions on neonics in Ontario are a result of the massive numbers of honeybees killed in that province because of the application of neonics on agricultural crops such as corn.[11]

doused with a neonicotinoid (a.k.a. neonic) that killed more than 50,000 bumblebees over the course of a few days—all because people didn't want honeydew from aphids dripping onto parked cars. This massive death was given wide publicity, bringing attention to the crisis-level loss of so many bees from coast to coast across North America.

In addition to the direct contact bees have with lethal levels of the toxins when they forage on nectar and pollen contaminated by neonics, the dust used to coat the seeds gets in the air and can end up on bees.[12] In fact, the very machines used to plant the seeds spread the toxins in the dust emitted by their exhaust. And honeybees drink water in puddles contaminated with neonics. At the very least, if neonics don't kill the bees on contact, they impair their foraging skills and affect their ability to navigate to and from the flower and the hive. Somewhere between 5,000 and 10,000 times more toxic than DDT, according to Dr. Jean-Marc Bonmatin of the National Centre for Scientific Research in France, neonics are so

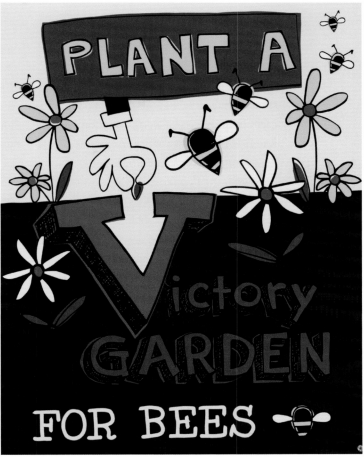

Clockwise from top left: A British Ministry of Agriculture poster produced in 1944 with artwork by Le Bon (IMAGE COURTESY THE NATIONAL ARCHIVES). Folks with Victory Gardens for Bees can put up signs to declare their commitment to the cause (LYSA MAIR DESIGN). Joe Wirtheim's eye-catching signs celebrate the Victory Gardens of the present, past and future. (JOE WIRTHEIM PHOTO).

toxic that if a bird eats a couple of treated seeds it will likely die.[13]

Most corn crops produced in Canada and the US contain neonics, meaning that this neurotoxin is probably present in every part of North American–grown corn, along with many soy, wheat or canola plants unless grown organically. In addition to food crops, flowers raised for the floral industry are affected. Big-box stores, flower shops and nurseries generally demand pristine plants with no sign of aphids or caterpillars. To ensure this, neonics are used to coat seeds or applied to the soil. As crops grow, they absorb the toxin, making plants poisonous to any insect that nibbles the roots, stem, leaves, pollen, nectar, sap or even dead foliage. And in addition to poisoning bees, neonics are in the food we eat and often the water we drink. These toxins remain in the soil and woody stems of plants for months or years after one application. Contaminated soil poisons future generations of plants—and bees, and us.

Dr. Elle suspects bumblebees in her own bee garden

Sound artist Peter Courtemanche records bees in Oak Meadows Park in Vancouver. The biodiversity of a garden is reflected in the complex layers of sound produced by its pollinators.

may have been poisoned by neonic-treated salvia. "I bought sage for the bees in my garden, but saw some clearly dying of pesticide exposure, trembling and walking in circles on the ground. The next day, I read a press release that stated some annual plants were tested for neonics and that salvia had high levels in the plants."[14] When you are buying plants for your garden, phone the store ahead of time to talk to someone on staff who can answer this question knowledgeably and truthfully: "Can you guarantee the plants I buy at your store have not been treated with neonics at any time in their growth cycle?"

Because of pressure from bee activists, the use of neonics was partially banned in the European Union in 2013. (They have been banned on sunflowers and maize in France since 2004, with no resulting loss of crop yields.[15]) Neonics are still legal in Canada and the US, but there is growing protest against their use. Vancouver became the first Canadian city to ban neonics in city parks through the park board's Pollinator Project initiative in October 2014, and Montreal followed suit in May 2015. On July 1, 2015, the Ontario government was the first in North America to restrict the use of seeds treated with neonics.[16] The US Fish

and Wildlife Service issued a complete ban as of
January 2016 on the use of these bee-killing toxins on
its 150 million acres of wildlife refuge.[17] It appears public pressure and new scientific evidence are finally starting to turn the tide against these bee-killing chemicals.

Growing a robust garden without the use of chemicals is at the heart of creating a Victory Garden for Bees. Avoid conventional herbicides, pesticides, miticides and fungicides, all which kill or seriously harm bees. Even some biological controls can be harmful for bees, so do your research before using them. Herbicides and pesticides that seep into the ground or accumulate in the soil will destroy ground-nesting bees and poison their habitat. Herbicide and pesticide residue on leaves and in water, pollen and nectar will kill any kind of bee. Instead of herbicides, pull weeds by hand or torch them where and when it is not going to be a fire hazard. Grow plants in communities so that they grow together tightly, leaving less opportunity for weeds to pop up. Use biodiversity, companion planting, beetle banks and insectaries to keep your garden healthy and balanced, and toxins away from you, your family and all bees.

2. EMBRACE VARIETY

Essential to successful organic gardening is increasing the number and variety of plants that attract pollinators and other beneficial insects to your garden. Everything you do to strengthen biodiversity will also make your garden more hospitable for bees. Increase productivity and blossom density by planting vines and plants that can grow up rather than out: California honeysuckle, clematis, kiwi, runner beans and more. Plant flowers that clamber over bushes to bloom. Consider adding bee gardens on the rooftops of sheds, playhouses and garages. Erect vertical gardens with such bee treats as strawberries and sedum.

The key is balancing volume with variety and succession of bloom so that you have a crucial number of flowers available at any given time during the active bee season, with enough diversity of blossoms to support the maximum variety of bee species. The best strategy to boost beauty and biodiversity is to create a bee garden that stimulates your senses with colour, light, scent, texture, taste and bee sounds. Add

Spurring on Bees

Canada columbine (*Aquilegia canadensis*) evolved with long, red nectar spurs because it is pollinated by hummingbirds, which are attracted to red, tubular flowers. In England, where they don't have hummingbirds, the flowers have shorter nectar spurs, designed for bees. They are also typically blue, purple and yellow—all colours with great appeal for bumblebees.

bee-supportive herbs, native plants, trees, shrubs, vegetables and vines to custom fit your garden to feed bees as you feed your family, body and soul.

3. PLANT A SUCCESSION OF BLOOM

As well as having a variety of plants flowering simultaneously, a bee garden should travel well through time. A succession of plants with overlapping periods of bloom will ensure that there will be no gaps in your garden's ability to sustain bees. And this is critical because, while honeybees can store large amounts of food, many other bees cannot.

Think about breaking each of the spring, summer and fall seasons into the three flowering windows:

Opposite, clockwise from top left: After forget-me-nots are pollinated, the centres turn from yellow to white. UK artist Rebecca Chesney installed her meadow project "I'm blue, you're yellow" in Liverpool's Everton Park. A yellow-faced bumblebee (*Bombus vosnesenskii*) follows the nectar guides of an obedience plant; some nectar guides are seen by bees, but are invisible to humans. Five spot flowers have dramatic dark purple spots.

early, middle and late. Blooms for bees are most important from early spring, when the first bumblebee queens emerge, to mid-fall, when they enter their hibernacula. Use long-blooming perennials and annuals to help plug any gaps left by plants with a shorter bloom period.

4. SHAPE RELATIONSHIPS WITH BEES

When talking about the form of the flower and its relationship to the pollinating bee, we mean the shape of the nectar and pollen delivery system. As a bee gardener, learn to look at a plant and ask two key questions: how deep is the nectar tube of the flower or the florets, and how long is the proboscis (tongue) of the bee I hope to support? Flowers can have a radial (circular) symmetry or consist of two halves that mirror each other. The nectar tubes can be tiny and shallow, as in the case of the forget-me-nots that support tiny carpenter bees, or large and deep, like the red trumpet honeysuckle loved by large bumblebee queens.

If the nectar fills a flower to a certain point, shorter-tongued bees can sip it up, using their tongues like straws in a tall milkshake glass, but they do best with a shallow cup. And while long-tongued bees can sip nectar from shallow flowers, it is more efficient to drink from the deeper flowers that suit the length of their proboscises. So make sure you have an abundance of nectar-producing flowers for bees of all tongue lengths, with an array of shallow, medium and deep bowls.

5. PLAN PLANT TEAMWORK

Don't treat your yard like one homogenous zone—break it down into microclimates to fill your garden with the optimum variety of plants to attract bees and the other insects that work to make your flowers and fruits healthy and productive. And instead of plunking in a plant here and there, plant in clusters—larger groupings of flowers attract more bees.

Bees (and butterflies) like groupings of plants at least 3 by 3 feet (1 square metre). Grow plants in groups of 3 to 5 in order to optimize pollination and provide some stability and aesthetic appeal.

6. CONSIDER POLLEN AND NECTAR PAYBACK

Evaluate the ecological function of each plant in your garden. Does it act as architecture but lack food for bees? Is the plant as purely ornamental as a plastic pink flamingo, or does it offer bees breakfast, lunch and dinner? As you evaluate the benefits of the herbs and vegetables in your garden, consider the nutritional payback they provide to bees. Seek flowers that provide protein-rich pollen and sugar-rich nectar for your garden's precious pollinators.

7. CHOOSE SUNNY SPOTS

Any part of your garden that gets six or more hours of sunlight is prime real estate for bees and bee plants.

Opposite, clockwise from top left: A sweat bee (*Lasioglossum*) forages on yarrow. A honeybee is covered in blue-black pollen from oriental poppies. A cheeky wool carder bee (*Anthidium manicatum*) sticks its tongue out for the camera. Lupin blossoms have a heavy bottom lip that can only be tripped by larger bees.

Bees love sun-warmed nectar and also need sunny spots to warm up their bodies so they can get to work.

Start with the extremes, which you're probably aware of already: which is the hottest and sunniest part of your garden, and which area gets the most shade? The warmest spot may work well as a Mediterranean zone where rosemary, thyme, oregano and other heat-loving herbs will thrive.

A shady northeastern side, on the other hand, can host native rainforest plants with salmonberry and thimbleberry bushes, or other low-light, bee-friendly themes. Bumblebees in particular have a tolerance for foraging in the shade because of their furry coats and ability to thermoregulate, so focus shadier areas on bumblebee plants.

8. LET SOME VEGETABLES AND HERBS BOLT

Plant enough vegetables and herbs that you can afford to allow some to selectively bolt. The blooms of plants in the cabbage, carrot and mint families are particularly helpful to bees. You can cut off buds as needed to hold back some plants from blooming until you feel it's time to let them go for it. A bonus to letting kale, parsley, cilantro and other plants bolt and bloom is that if you leave them in long enough, you will be rewarded with a windfall of seeds for next season's plantings.

9. COLOUR YOUR GARDEN BEE-BEAUTIFUL

Yellow bee flowers bring the sunlight down to earth. Many bee-pollinated plants have yellow centres and/or pollen. Some yellow flowers, such as brown-eyed Susan, reveal secret bee patterns under ultraviolet light—such as a marked contrast between a dark centre and the light outer edges of the petals—that are ordinarily invisible to the human eye. Since bees are better at seeing contrast than colour, patterns trump tints. Although many bee plants are yellow, white, blue and purple, there are many flowers and patterns that bees see as "bee purple" (see A Peace Garden for Bees on page 186). Bumblebees in particular also forage at many of the same red flowers that attract humming-birds. White and pink bee-attracting flowers often have red-violet nectar guides in the form of lines or spots. Think of the speckled throats of foxglove blossoms or the lines inside white-flowering penstemon. These markings act like neon signs, drawing the bees in to the local diner.

10. BEE WATER WISE

While conserving water is good for man and bee, starving nectar-bearing plants of hydration is not recommended. When plants are blooming, they need to have their basic needs met so they can use their energy to produce an abundance of nectar. Insects with shorter tongues are particularly vulnerable as they need flowers to have a high nectar level so they can use their tongues to sip up the top layer. If you have nectar-bearing plants like Dutch clover in your lawn, rather than letting the lawn go completely dormant in the dry months, at least focus the watering on the patches with the most flowers. Although it's time-consuming, saving grey water for your clover is a water-wise way to support bees. It will also help the earthworms and microbes in your soil.

Providing water sources in your garden is essential for increasing pollinator biodiversity. Create a rehydration station for bees to use from Valentine's Day to Halloween. Place a dish of clean water with moss, rocks and branches for bees to perch on while sipping

Opposite, clockwise from top left: A leafcutter bee gathers pollen from a tidy tips flower (JACK TUPPER PHOTO). A masked bee sips nectar from the "yolk" of a poached egg flower. Create a bee oasis from an artfully arranged collection of natural objects. Shrubby cinquefoil supports many species of bees, including this sweat bee (*Lasioglossum*).

The Good News about European Chafer Beetles

As European chafer beetles and their predators ravage lawns in the neighbourhoods around me, I am silently cheering them on. Skunks, raccoons, flickers and crows are Vancouver foodies that love the new culinary trend of raw exotic larvae. And the scruffier the lawn, the more it appeals to little ground-nesting bees. On a sunny day, you can look at lawns like this and see masses of them digging holes and laying eggs, with cuckoo bees lurking and nipping in to lay their eggs and then shirking their maternal duties. Municipalities in areas stricken by infestations are recommending Dutch clover as a replacement for damaged lawns. One of my neighbours noted, "We replaced our turf with clover and now our lawn is buzzing with bees."

"We have herbicides and power mowers, so we're going to use them. We think, 'Wouldn't it be great if the entire world looked like picture-book houses?' And that kind of environment is just a train wreck for animals."[21]

—SAM DROEGE, head of the US Geological Survey's Bee Inventory and Monitoring Lab

so they don't fall in and drown. Entomologist and gardener Eric Grissell suggests creating a miniature solar-powered fountain which will provide water for a wide variety of insects, including bees and butterflies.[20] Bees will also sip water droplets that pool on leaves.

11. SLOW MOW OR LET THE CLOVER GROW

Try to raise the level on the mower to avoid beheading flowers in the lawn. Wait until just before the flowers go to seed to cut, and mow during the day to protect the bees that are sleeping in the blossoms at dusk. Alternate areas of mown and unmown zones so there is always some forage available for bees. When making your lawn bee-friendly, remind family and friends to wear shoes to protect their feet from bee stings when treading on the turf.

12. DO NOT DISTURB THE DIRT

Since 70 percent of bees nest in the ground, digging and tilling can have a hugely negative impact on bee habitat. Try to become aware of ground-nesting sites and leave some parts of your garden untouched—that's where bees will nest. Also, cultivate areas of perennial crops and flowers to cut down on tillage in those areas. When choosing crop rotation and weeding practices, look for low-till or no-till options. And try to include different exposures and soil types to attract a variety of ground-nesting bees that have unique soil preferences for habitat.

13. PLAN PRUNING AND PLANT DIVISION

Dividing of perennials happens in spring and fall: early spring is a marginally better time to move these plants to avoid disturbing undeveloped bees or newly laid eggs. Rotate areas of garden disturbance in places with ground-nesting bees. If you must dig in areas with nest sites, make sure you leave some sections alone for the populations to regenerate and bounce back. Be aware of bees nesting in any dried stems.

14. MINIMIZE MULCHING

Plastic barriers and landscaping cloth will smother ground-nesting bees. Particularly avoid using this in fields where there are squash bees. Natural mulch can be used judiciously. Avoid any dyed materials. A light layer of leaves left to decompose naturally is best, as emerging bees will be able to dig up through it or down into a nest. Avoid mulching at all if possible during high-volume nesting seasons between late spring and early fall or during bumblebee queen hibernation in the fall to early spring.

15. LEAVE WEEDS

News flash for busy folks and "wanna-be-lazy" gardeners: gardens without weeds are less attractive to bees. Naturally, anything given an official designation as invasive or noxious should be removed and replanted with better bee forage, preferably a native species, but many weeds are simply medicinal or edible plants brought by yesterday's pioneers that have happily naturalized in North America.

16. TAKE ADVANTAGE OF COMPANION PLANTING

Many of the plants that feed bees also attract carnivorous insects that feed on the predators that want to eat your plants. These beneficial insects include a motley crew of parasitic wasps, syrphid flies, lacewings and beetles. The adults and larvae of these insects help to keep the garden in balance, so the aim of the eco-sensitive gardener is not to destroy every "bad bug" in the garden, because beneficial insects and hungry birds need to eat too. When planting, aim for Elysian Fields of plant and insect diversity, not the killing fields of monoculture and pesticides. Your garden should be a symphony of insect sounds—alive with sonic diversity.

It doesn't take much space to provide a bit of food and shelter for backyard bees. Just make sure the nesting stems are tightly packed together, firmly held in place and protected from rain.

"WEEDS" TO LEAVE for Bees

A SEASONAL PLANTING CHART

*I*n some cases, plants we classify as unwanted weeds are important sources of food that bees depend on for their survival—and some are even valuable to us. As long as they are not invasive to the region you are gardening in, there are weeds worth leaving to at least produce flowers, if not seeds. Some gardeners will be happy to learn that neglected patches in their lawns and gardens can now be justified by putting up a little sign that says "Reserved for VIBS—Very Important Bees."

Edible weeds should be consumed in moderation. In particular, pregnant women and those with specific health concerns are advised to check with a medical doctor prior to consuming them.

PLANT ORIGIN

OW: Old World (indigenous to Europe, Asia or Africa)

NW: New World (indigenous to the Americas and their islands)

BLOOM PERIOD

E: Early

M: Mid-season

L: Late

SUC: May be seeded in succession

HARDINESS

refers to the coldest zone the plant is hardy to.

DEER AND DROUGHT RESISTANCE are most reliable once the plant is established.

BEES ATTRACTED

BB: Bumblebees

HB: Honeybees

SB: Solitary bees

BI: Beneficial insects

PLANT NAME	BLOOM AND PLANT FAMILY	HARDINESS	HEIGHT AND PLANT NOTES	BENEFITS TO BEES*
Cat's Ear, a.k.a. False Dandelion *Hypochaeris radicata* **OW** Noxious in WA	**M–L** Aster. Yellow flowers resemble dandelion, but smaller.	**Perennial** Hardy to zone 3.	**4–24 in (10–60 cm)** Leaves covered in fine hairs, lack jagged edges of dandelion and stems are thinner. Young leaves are edible.	**BB, HB, SB, BI** Mining and sweat bees. Particularly important to short-tongued bees.
Clover, Dutch *Trifolium repens* **OW**	**M–L** Legume. White blossoms, some with pink tinge.	**Perennial** Hardy to zone 3.	**4–6 in (10–15 cm)** Common in lawns. Young leaves and flowers edible. Tea from dried flowers used to treat colds.	**BB, HB, SB, BI** Cuckoo and mason bees. Top honey plant in North America.
Clover, Red *Trifolium pratense* **OW**	**M–L** Legume. Flowers are more pinky-purple than red.	**Perennial** Hardy to zone 3.	**8–30 in (20–76 cm)** Mix with Dutch clover for forage bumblebees don't have to compete with honeybees for. Tea from flowers used to combat cancer and gastric and bronchial ailments.	**BB, SB, BI** Digger, large leafcutter and mason bees. Not the best for honeybees because corolla tubes are too long.
Comfrey, Common *Symphytum officinale* **OW**	**E–M** Borage. Pink, mauve and white bell-shaped blossoms.	**Perennial** Hardy to zone 3.	**Up to 80 in (2 m)** Likes well-drained soil. Excellent soil amendment when used to make compost or compost tea. Useful as topical compress for bruises.	**BB, SB** Buzz-pollinated, bumblebee favourite. Blossoms accessed by shorter-tongued bees nectar robbing. Refills with nectar every 45 minutes.
Daisy, English *Bellis perennis* **OW**	**E–M–L** Aster. Classic small daisy with white petals and yellow centre. Deadhead.	**Biennial** Hardy to zone 6.	**Less than 6 in (15 cm)** Used as topical treatment in salves for skin bruising and inflammation; internally for digestive ills. Young leaves and petals edible; pickle buds.	**SB, BI** Cuckoo, mining, small carpenter and sweat bees.
Dandelion *Taraxacum officinale* **OW**	**E–M–L** Aster. Yellow flower rays cover entire circular blossom.	**Perennial** Hardy to zone 3.	**4–12 in (10–30 cm)** An invasive but very useful herb. All parts of the plant are medicinal, used to treat liver, kidney and stomach ailments. Young leaves can be eaten raw and petals used to make wine.	**HB, SB, BI** Orange pollen. Mason, mining, sweat bees. Honey potential: 101–200 lbs/ac. Sugar 74 percent. Important early source of nectar for bees during brood rearing.
Deadnettle, Henbit *Lamium amplexicaule* **OW**	**E** Mint. Tiny purple bilabial blossoms with flower tubes.	**Annual**	**4–12 in (10–30 cm)** Commonly seen blanketing disturbed farmland. Flowers pink to purple. Young leaves eaten raw or as a pot herb.	**BB, HB, SB, BI** Long-tongued bees. Key for new spring bumblebee queens. Honeybees are nectar robbers.
Goldenrod, Grey *Solidago nemoralis* **NW**	**M–L** Aster. Spikes of small, shallow golden flowers. Silvery foliage.	**Perennial** Hardy to zone 2.	**8–40 in (20–100 cm)** Pioneer species. Can become weedy and aggressive in moist, fertile fields. A salve made from the flowers soothes bee stings.	**BB, HB, SB, BI** Mining, large carpenter, sweat bees and more. Honey plant.
Miner's Lettuce *Claytonia sibirica* and *C. perfoliata* **NW/OW**	**E–M–L** Purslane. Tiny daisy-like flowers with white or pink petals.	**Annual/ Perennial** Hardy to zone 8.	**0.4–15 in (1–38 cm)** Found in shady, moist, well-drained sites. Edible leaves can be harvested from spring to autumn. Rich in vitamin C.	**HB, SB** Visited by smaller bees, mostly for pollen.

*Each pound per acre is equal to approximately 1.1 kilograms per hectare.

PLANT NAME	BLOOM AND PLANT FAMILY	HARDINESS	HEIGHT AND PLANT NOTES	BENEFITS TO BEES*
Smartweed, a.k.a. Lady's Thumb *Polygonum persicaria* **OW**	E–M–L Buckwheat. Creeping branched stems with clusters of pink or white blossoms.	Annual	8–40 in (20–100 cm) Many species of *Polygonum* have naturalized in North America. Skilled herbalists use plant as a diuretic blood coagulant. Use sparingly to season foods; peppery flavour.	BB, HB, SB, BI Sweat bee. Honey plant. The cultivated varieties of *P. persicaria* are better for bees, but weeds are a good supplement.
Sorrel, Common Wood *Oxalis stricta* **NW**	M–L Wood sorrel. Delicate yellow blossoms. Clover-like leaf.	Annual/ Perennial Hardy to zone 3.	6–20 in (15–50 cm) Edible in small amounts. Add to salads and infuse leaves and flowers for lemon flavour. High in vitamin C. Dye plant.	SB, BI Green sweat, small carpenter and sweat bees.
Willowherb, Fringed *Epilobium ciliatum* **NW**	M–L Primrose. Delicate purple flowers.	Perennial Hardy to zone 3.	12–40 in (30–100 cm) This native plant is easily pulled before going to seed in small plots but can be egregious in farms or nurseries.	SB Small carpenter and sweat bees. Specialist pollinator *Lasioglossum oenothera*. Leaf used by leafcutter bee.
Yarrow, White *Achillea millefolium* **NW**	M–L Aster. Flat clusters of small white flowers. Ferny foliage.	Perennial Hardy to zone 3.	20–24 in (50–60 cm) Spreads by rhizomes and seeds. Naturalizes into lawns. Ferny foliage can staunch a bloody nose or small cut.	BB, HB, SB, BI Mining and sweat bees. Wool carder bees gather hairs along stems. Essential gap-filler for short-tongued bees.

*Each pound per acre is equal to approximately 1.1 kilograms per hectare.

The yellow blossoms of cat's ear (a.k.a. false dandelion) provide important sustenance for bees of all stripes. Creating no-mow zones allows these bee-feeding weeds to flourish.

Bee plants WITH BENEFITS

A COMPANION PLANTING CHART

*W*hen planning your garden, choose a variety of companion plants that bloom in succession through the seasons. Be sure to include a diverse selection of five of the key power companion-plant families: allium, aster, borage, mint and carrot (and see the chart for more ideas). Choosing plants that bloom for a long time will increase their ability to attract bees and other beneficial insects. Highly aromatic plants also help confuse pests and deter them from finding their favourite foods. Many of these companion plants are culinary or medicinal herbs, so the benefits of planting them are bountiful and varied.

PLANT ORIGIN

OW: Old World (indigenous to Europe, Asia or Africa)

NW: New World (indigenous to the Americas and their islands)

BLOOM PERIOD

E: Early

M: Mid-season

L: Late

SUC: May be seeded in succession

BEES ATTRACTED

BB: Bumblebees

HB: Honeybees

SB: Solitary bees

BI: Beneficial insects

PLANT NAME	BLOOM TIME, PLANT FAMILY AND COMPANION-GARDENING NOTES	BEES THAT BENEFIT FROM BLOOMS	OTHER BENEFICIAL INSECTS ATTRACTED
Alyssum, Sweet *Lobularia maritima*	**M** Brassica. General companion. Long-blooming.	**SB** Sweat bees.	Lacewings, parasitic wasps, syrphid flies, tachinid flies.
Anise *Pimpinella anisum*	**M–L** Carrot. Masks odour of brassica to deter enemies. Friend of cilantro. Improves health of nearby plants. Repels aphids.	**HB, SB, BI** Scent attracts bees. Crushed seeds in honey used as a lure for honeybee swarms in bait hives.	Lacewings, ladybugs, parasitic wasps, tachinid flies.
Anise Hyssop *Agastache foeniculum*	**M–L** Mint. Plant by brassica and plants that draw aphids. Attracts predators of cabbage butterfly, cabbage looper. Avoid radish.	**BB, HB, SB** Exceptional honey plant. Digger, leafcutter, masked, mining bees.	Parasitic wasps, syrphid flies.
Aster *Symphyotrichum* spp. cultivars and native species	**M–L** Aster. Plant by asparagus. Good general companion.	**BB, HB, SB, BI** Important fall source of pollen, nectar for native bees.	Damsel bugs, ladybugs, minute pirate bugs, parasitic wasps, soldier beetles, tachinid flies.
Basil *Ocimum* spp.	**M–L** Mint. Plant by tomato, pepper. Avoid rue. Repels aphids, asparagus beetle, flies, mosquitoes, spider mite, thrips.	**BB, HB, SB** Basil attracts bumblebees that pollinate pepper, tomato. Honey plant.	Butterflies.
Bee Balm, Scarlet *Monarda didyma*	**M–L** Mint. Improves flavour, growth of tomatoes.	**BB, SB** Elongated flowers loved by bumblebee guilds, queens.	Syrphid flies. Attracts hummingbirds (which gobble up whiteflies).
Bee Balm, Spotted *Monarda punctata*	**M–L** Mint. Plant by cabbage, potato, tomato. Repels chiggers, fleas, mosquitoes. Nectar plant for endangered Karner blue butterflies.	**BB, HB, SB, BI** Long-horned, mining, plasterer, sweat bees. The most accessible *Monarda* for bees.	Braconid, chalcid and cynipid wasps; hummingbirds; ladybugs; minute pirate bugs; predaceous ground beetles; soldier beetles.
Borage *Borago officinalis*	**E–M** Borage. Plant by strawberry, squash, tomato. Soil amender. Repels tomato hornworm.	**BB, HB** Honey plant. Refills with nectar every 2 minutes.	Lacewings, parasitic wasps.
Brassicas *Brassica* spp. Broccoli, Brussels sprouts, kale, mustard, turnip	**E–M** Brassica. Plant by beet, celery, chamomile, chard, dill, lavender, mint, onion, pennyroyal, rosemary, sage, thyme. Mustard used as decoy plant for cabbage butterfly.	**BB, HB, SB, BI** Mason bees and more. High amounts of pollen and nectar. Many brassicas are good honey plants.	Ladybugs, parasitic wasps, syrphid flies.
Brown-eyed Susan *Rudbeckia* spp.	**M–L** Aster. General long-blooming companion that beautifies the garden and attracts a wealth of pollinators and predatory helpers. A fast-growing plant suitable for restoration and guerilla gardening projects.	**BB, HB, SB, BI** Cuckoo, green sweat, leafcutter, long-horned, small carpenter, sweat bees. Specialist pollinators *Andrena rudbeckia*, *Heterosarus rudbeckia*.	Bee flies, parasitic wasps, soldier beetles, syrphid flies, tachinid flies.
Buckwheat *Fagopyrum esculentum*	**M** Buckwheat. Plant near squash, bell pepper. General powerful companion plant. Soil amender (planted as a green manure and soil "recharger").	**BB, HB, SB, BI** Honey plant.	Assassin bugs, big-eyed bugs, damsel bugs, ladybugs, minute pirate bugs, parasitic wasps, predaceous beetles, spiders, spined soldier bugs, syrphid flies, tachinid flies.
Bunchberry *Cornus canadensis*	**M** Dogwood. General companion, especially as orchard or hedgerow understory.	**BB, HB, SB, BI** Cuckoo bumblebee, cuckoo, mining, sweat bees.	Parasitic wasps, syrphid flies.

PLANT NAME	BLOOM TIME, PLANT FAMILY AND COMPANION-GARDENING NOTES	BEES THAT BENEFIT FROM BLOOMS	OTHER BENEFICIAL INSECTS ATTRACTED
Calendula *Calendula officinalis*	**E–M–L** Aster. Plant near asparagus, tomato. Long-blooming general companion. Repels asparagus beetle, tomato hornworm, some nematodes.	**BB, HB, SB, BI** Cuckoo, metallic green, leaf-cutter, mason, small carpenter, sweat, wool carder bees.	Ladybugs, parasite wasps, syrphid flies.
California Bluebell *Phacelia campanularia*	**E–M** Borage. General companion plant. Attracts aphid predators.	**BB, HB, SB** Mason, mining bees.	Syrphid flies.
Carrot *Daucus carota*	**M** Carrot. Plant by chives, lettuce, leek, onion, rosemary, sage, tomato. Umbels act as landing pads and travel corridors for bees and other insects.	**BB, HB, SB** Overwinter for spring blossoms for cuckoo, leafcutter, masked, mining, sweat bees.	Assassin bugs, big-eyed bugs, lacewings, parasitic wasps.
Catnip *Nepeta* spp.	**M–L** Mint. Plant by eggplant. Repels ants, aphids, flea beetle, Japanese beetle, squash bug, weevil.	**BB, HB, SB** Cuckoo, digger, leafcutter, sweat bees. Honey plant.	Bee flies, parasitic wasps, soldier bugs, syrphid flies, tachinid flies.
Chives *Allium schoenoprasum*	**E–M** Amaryllis. Plant by apple, brassica, carrot, grape, tomato. Repels aphids, Japanese beetle.	**BB, SB, HB, BI** Digger, green sweat, sweat bees.	Syrphid flies.
Cinquefoil, Shrubby, a.k.a. Potentilla *Dasiphora fruticosa*	**M–L** Rose. Good bones for a bee garden. Powerhouse bee and beneficial-insect plant. Groups well with small *Spiraea* shrubs, salvia, catmint. Repels nematodes.	**BB, HB, SB** Masked, sweat bees.	Asassin bugs, big-eyed bugs, damsel bugs, green lacewings, ladybugs, minute pirate bugs, parasitic wasps, spined soldier bugs, syrphid flies.
Clover, Crimson *Trifolium incarnatum*	**E–M–L–SUC** Legume. Plant by brassica, grape. Amends soil. Repels cabbage aphid, cabbage worms.	**BB, SB, HB** Digger, large leafcutter, mason bees.	Big-eyed bugs, minute pirate bugs.
Clover, Dutch *Trifolium repens*	**E–M–L–SUC** Legume. Plant near brassica, grape. Soil amender. Repels cabbage aphid, cabbage worms.	**BB, HB, SB** Digger, large leafcutter, mason bees.	Aphid midges, big-eyed bugs, ladybugs, minute pirate bugs, parasitic wasps, predatory ground beetles, tachinid flies.
Comfrey *Symphytum officinale*	**E–M** Borage. Plant near fruit trees. Soil amender. Trap plant for slugs.	**BB, SB** Buzz-pollinated bumblebee favourite. Nectar robbing by shorter-tongued bees.	Lacewings, parasitic wasps.
Coneflower, Yellow *Ratibida pinnata*	**M–L** Aster. General companion. Long-blooming essential native bee plant. Seeds heads have an anise fragrance when crushed.	**BB, HB, SB, BI** Cuckoo, leafcutter, long-horned, sweat bees. Specialist bee *Andrena rudbeckia*.	Chalcid wasps, ichneumon wasps, lacewings, ladybugs, minute pirate bugs, soldier beetles, spiders, syrphid flies, tachinid flies.
Coriander, a.k.a. Cilantro *Coriandrum sativum*	**M–L–SUC** Carrot. Plant near anise, brassica, caraway, carrot, dill, potato, spinach. Avoid fennel. Make into a spray for potato beetle, spider mite.	**HB, SB, BI** Mining, plasterer, sweat bees. Varroa mites are repelled by the essential oil of the seeds. Honey plant.	Lacewings, ladybugs, minute pirate bugs, parasitic wasps, soldier beetles, syrphid flies, tachinid flies.
Cosmos *Cosmos* spp.	**M–L** Aster. General garden companion. Airy fringed foliage works well with veggies. Repels Mexican bean beetle. Try vanilla-scented chocolate cosmos for fun.	**BB, HB, SB** Cuckoo, green sweat, leafcutter, long-horned, mason, mining, sunflower, sweat bees. Petals used by leafcutters.	Big-eyed bugs, damsel bugs, lacewings, ladybugs, minute pirate bugs, spiders, syrphid flies, tachinid flies.
Cucumbers *Cucumis sativus*	**M** Squash. Plant near bean, corn, dill, nasturtium. Avoid sage, potato, rue.	**BB, HB, SB** Specialists include squash bees.	Syrphid flies.

PLANT NAME	BLOOM TIME, PLANT FAMILY AND COMPANION-GARDENING NOTES	BEES THAT BENEFIT FROM BLOOMS	OTHER BENEFICIAL INSECTS ATTRACTED
Cup Plant *Silphium perfoliatum*	L Aster. General companion plant. Plant by bergamot, sea holly, milkweed, coneflower for companion powerhouse.	BB, HB, SB, BI Cuckoo, digger, green sweat, leafcutter, long-horned, mining, small carpenter bees.	Lacewings, ladybugs, minute pirate bugs, soldier bugs, spiders.
Dahlia *Dahlia* spp.	M–L Aster. General companion. Repels nematodes.	BB, HB, SB, BI Late-season gap filler for variety of bees.	Syrphid flies.
Dead Nettle, Henbit *Lamium amplexicaule* and **Dead Nettle, Red** *Lamium purpureum*	E Mint. Plant by beet, carrot, pea, radish. Deters potato bug, other pests. General insect repellant.	HB, SB Long-tongued bees access nectar. Key for new spring bumblebee queens. Honeybees benefit from nectar robbing.	Bee flies. Butterfly plant.
Dill *Anethum graveolens*	M Carrot. Plant by cabbage, corn, cucumber, lettuce, onion. Avoid caraway, carrot, lavender, tomato. Repels aphids, cabbage looper, spider mite, squash bug, tomato hookworm.	HB, SB Small carpenter, sweat bees. Single variety honey plant in Italy. Provides nectar and pollen.	Damsel bugs, ladybugs, parasitic wasps, syrphid flies. Host plant for swallowtail butterflies.
Elderberry *Sambucus* spp.	E–M Adoxa. Soil amender. Add leaves to compost. Deters aphids, carrot root fly, cucumber beetle, peach tree borers. Use for insect spray.	HB, SB Small carpenter, sweat bees. Small carpenter and mason bees use hollow stems for habitat. Important pollen plant.	Bee flies, syrphid flies, and many beetles, including long-horned, flower.
False Sunflower *Heliopsis* spp. Cultivars and native species	M–L Aster. General companion. Long-blooming bee magnet. One of the most reliable, hardy native plants for your garden.	BB, HB, SB, BI Cuckoo, green sweat, leafcutter, long-horned, mining, small carpenter, sweat bees.	Bee flies, lacewings, ladybugs, parasitic wasps, spined soldier bugs, syrphid flies, tachinid flies.
Fennel *Foeniculum vulgare*	M Carrot. Avoid most plants, as fennel inhibits growth of neighbouring plants. Repels aphids, fleas. Suppresses nearby weeds.	BB, HB, SB, BI Masked, mining bees. Honey plant. Stubble-cut and dry stems for tunnel-nesting bees.	Big-eyed bugs, ladybugs, parasitic wasps, pirate bugs damsel bugs, syrphid flies, tachinid flies. Host plant for caterpillars of swallowtail butterflies.
Flax *Linum* spp. Cultivars and native species	E–M–L–SUC Flax. Plant near carrot, potato. Deters Colorado potato beetle.	BB, HB, SB, BI Leafcutter, mining, sweat bees.	Butterfly larvae, syrphid flies.
Garlic *Allium sativum*	M Amaryllis. Orchard companion. Plant near celery, cucumber, lettuce, rose. Repels aphids, codling moth, Japanese beetle, nematodes, root maggot, snails. Make into a spray to kill aphids, fungus gnat, whitefly.	BB, HB, SB Let some hardneck garlic develop flower heads to provide nectar and pollen for bees.	Butterflies.
Golden Alexanders *Zizia aurea*	E–M Carrot. Plant near chives, leek, lettuce, onion, pea, rosemary, sage, tomato. An early-blooming native umbel important for short-tongued bees.	BB, SB, BI Cuckoo, green metallic, mining, masked, small carpenter, sweat bees. In some regions, links with mason bee nesting.	Butterflies, parasitic wasps, and more.
Goldenrod *Solidago* spp.	M–L Aster. Important late-season general companion plant. Attracts a high biodiversity of insects. Many native species and cultivars available. Avoid invasive seaside goldenrod.	BB, HB, SB, BI Honey plant. Large carpenter, mining, sweat bees. Several Andrenid (mining bee) specialist pollinators.	Assassin bugs, big-eyed bugs, damsel bugs, goldenrod soldier beetles, ladybugs, pirate bugs, parasitic wasps, soldier beetles, spiders, syrphid flies.

PLANT NAME	BLOOM TIME, PLANT FAMILY AND COMPANION-GARDENING NOTES	BEES THAT BENEFIT FROM BLOOMS	OTHER BENEFICIAL INSECTS ATTRACTED
Horehound *Marrubium vulgare*	M Mint. Plant by tomatoes, peppers to stimulate production. Repels grasshoppers. Hardy.	**BB, HB, SB** Honey plant. Leafcutter, small carpenter bees.	Parasitic wasps, syrphid, tachinid flies.
Hyssop *Hyssopus officinalis*	M-L Mint. Plant by cabbage, grape. Attracts enemies of cabbage butterfly, cabbage looper.	**BB, HB, SB** Honey plant. Funnel-shaped flowers support shorter-tongued bees.	Syrphid flies.
Joe-Pye Weed *Eupatorium* spp.	M-L Aster. Important late-season general companion plant.	**BB, HB, SB** Honey plant. Leafcutter, mining, small carpenter, sweat bees.	Damsel bugs, ladybugs, minute pirate bugs, parasitic wasps, tachinid flies.
Larkspur *Delphinium* spp.	E-M Buttercup. Orchard, rose garden companion. Kills Japanese beetle. Native NW species only.	**BB** Long-tongued bumblebees.	Bee flies and hummingbirds.
Lavender *Lavandula* spp.	M-L-SUC Mint. Orchard companion. Repels codling moth, fleas, whitefly.	**BB, HB, SB** Honey plant. Leafcutter, mason, wool carder bees.	Syrphid flies.
Leek *Allium ampeloprasum*	M Amaryllis. Orchard companion. Plant by carrot, celery, onion. Avoid legumes. Repels carrot fly.	**BB, HB, SB** Honey plant. Overwinter some leeks to bloom for bees.	Syrphid flies.
Lemon Balm *Melissa officinalis*	M-L Mint. General deterrent. Sprinkle leaves in garden to repel mosquitoes, squash bug.	**BB, HB, SB** Rubbed on honeybee hives to call bees back home.	Parasitic wasps, syrphid flies, tachinid flies.
Lovage *Levisticum officinale*	M Carrot. General beneficial for health, flavour of other plants. Protects from wind, sun.	**BB, HB, SB** Mining bees.	Ground beetles, lacewings, ladybugs, syphid flies, tachinid flies, parasitic wasps.
Marjoram *Origanum majorana*	M-L Mint. Improves yield and flavour of asparagus, bean, chives, eggplant, squash.	**BB, SB, HB,** Sweat bees. Nectar, pollen.	Butterflies, parasitic wasps.
Mint *Mentha* spp. Cultivars and native species.	M Mint. Plant by cabbage, potato, tomato. Mulch brassica with cuttings. Repels aphids, cabbage butterfly, flea beetle, mice.	**BB, SB, HB, BI** Honey plant. Leafcutter, resin, sweat bees.	Earthworms, parasitic wasps, soldier beetle, syrphid flies.
Mint, Mountain *Pycnanthemum* spp.	M-L Mint. Deters cabbage butterfly. One of the most popular northwest bee plants.	**BB, HB, SB, BI** Honey plant. Cuckoo, green sweat, long-horned, masked, resin, sweat bees.	Beetles, butterflies, lacewings, parasitic wasps.
Nasturtium *Tropaeolum* spp.	M-L Nasturtium. Plant near cabbage, fruit trees, radish, squash, tomato. Repels squash bug, striped cucumber beetle, whitefly, woolly aphids. Trap plant for black aphids.	**BB, HB, SB** Very good for long-tongued bees; smaller bees lap up nectar at the edges of nectar spurs.	Loved by hummingbirds, which eat whiteflies.
Onion *Allium* spp. Cultivars and native species.	E-M-L-SUC Amaryllis. Orchard companions. Plant by beet, brassica, carrot, celery, cucumber, lettuce, parsnip, pepper, savory, spinach, squash, strawberry, tomato, turnip. Repels aphids, carrot root fly, cutworm, moles, slugs, weevils.	**BB, HB, SB, BI** Honey plant. Flower heads can contain hundreds of florets with nectar and pollen, which save bees energy and travel time.	Syrphid flies.
Oregano *Origanum* spp.	M-L Mint. Plant by brassica, cucumber, grape. Good overall companion. Deters cabbage butterfly, cucumber beetle.	**BB, SB, HB, BI** Long-horned, mason, sweat and more. Honey plant. High sugar, at 76 percent.	Big-eyed bugs, butterflies, minute pirate bugs, parasitic wasps, lacewing larvae, soldier beetles, syrphid flies, tachinid flies.

PLANT NAME	BLOOM TIME, PLANT FAMILY AND COMPANION-GARDENING NOTES	BEES THAT BENEFIT FROM BLOOMS	OTHER BENEFICIAL INSECTS ATTRACTED
Oregon Tea *Clinopodium douglasii*	M–L Mint. General companion plant. Repels slugs.	BB, SB Long-tongued bees seek nectar.	Syrphid flies.
Parsnip *Pastinaca sativa*	E–M Carrot. Plant by bush bean, garlic, marigold, onion, pea, pepper, potato, radish, squash.	BB, HB, SB, BI Small-tongued bees.	Parasitic wasps, syrphid flies and more.
Phacelia, Lacey *Phacelia tanacetifolia*	Borage. General companion plant. Plant near anything that attracts aphids.	BB, HB, SB Mason bees and more. Top honey plant. Blue pollen.	Parasitic wasps, predatory true beetles, syrphid flies, tachinid flies.
Pincushion Flower, a.k.a. Scabiosa *Scabiosa* spp.	M–L Honeysuckle. General long-blooming companion plant.	BB, HB, SB Leafcutter, mining bees and more.	Parasitic wasps, syrphid flies.
Poached Egg Flower *Limnanthes douglasii*	E–M–SUC Limnanthes. Plant by tomato. Self-seeds to create excellent ground cover.	BB, HB, SB Mining bees and more. Specialists include *Andrena limnanthis*.	Syrphid flies.
Radish *Raphanus sativus*	E–M–L–SUC Brassica. Plant by bean, beet, carrot, cucumber, lettuce, melon, nasturtium, parsnip, pea, spinach, squash. Avoid hyssop. Chervil improves flavour of radishes.	BB, HB, SB, BI Honey plant. Small sweat bees. Plant every two weeks in early spring and leave some to bolt and bloom for bees.	Syrphid flies.
Rosemary *Rosmarinus officinalis*	E Mint. Plant by bean, broccoli, cabbage, carrot, hot pepper, sage. Avoid basil, potato, pumpkin. Place cuttings around carrot crowns. Repels bean beetle, cabbage butterfly, carrot fly, slugs, snails.	BB, HB, SB Long-horned, mason, mining, small carpenter, sweat, wool carder bees. Honey plant.	Butterflies.
Sage *Salvia* spp. Cultivars and native species.	M Mint. Plant by bean, cabbage, carrot, lettuce, pea, rosemary. Avoid cucumber. Repels cabbage butterfly, carrot fly, flea beetle, slugs.	BB, HB, SB Primarily visited by bumblebees, wool carder bees. Honey plant.	Butterflies, ground beetles.
Savory, Summer *Satureja hortensis*	M Mint. Plant by bean, onion, sweet potato. Repels bean beetle, black aphid, cabbage butterfly, sweet potato weevil.	BB, HB, SB Honey plant.	Butterflies.
Speedwell *Veronica spicata*	E–M Plantain. Long-blooming perennial, general companion.	BB, HB, SB Honey plant.	Lacewings, syrphid flies.
Spiraea *Spiraea* spp. Cultivars and native species.	M–L Rose. General companion plant, good bones for an insectary garden.	BB, HB, SB Small-tongued bees.	Assassin bugs, damsel bugs, ground beetles, ladybugs, minute pirate bugs, parasitic wasps.
Squash *Cucurbita* spp.	M Squash. Plant by bean, corn, cucumber, dill, melon, mint, nasturtium, onion, oregano, radish. Avoid potato.	BB, HB, SB Long-horned, squash bees.	Sometimes insect-eating tree frogs shelter in blossoms.
Sunflower *Helianthus annuus* Cultivars and native species.	M Aster. Plant by amaranth, bean, corn, cucumber, squash. Ants farm aphids on sunflower stalks; use as a decoy plant and remove with soapy water.	BB, HB, SB Digger, leafcutter, mining, sweat bees. Specialist bees visit sunflowers, including *Melissodes robustior*.	Big-eyed bugs, hyperparasitoid wasps, lacewings, parasitic wasps, predaceous stink bugs, robber flies, soldier beetles, spiders.

PLANT NAME	BLOOM TIME, PLANT FAMILY AND COMPANION-GARDENING NOTES	BEES THAT BENEFIT FROM BLOOMS	OTHER BENEFICIAL INSECTS ATTRACTED
Thyme *Thymus* spp.	**M** Mint. General companion. Plant near cabbage to repel cabbage worm, whitefly. Excellent "living mulch" around fruit trees.	**BB, SB, HB** Carder, cuckoo, leafcutter, mason, mining, resin, small carpenter, sweat bees.	Parasitic wasps, syrphid flies, tachinid flies.
Tickseed, a.k.a Coreopsis *Coreopsis* spp.	**M–L** Aster. General companion plant. Long-blooming with multiple flower heads.	**BB, HB, SB, BI** Cuckoo, leafcutter, long-horned, small carpenter, small resin bees.	Lacewings, minute pirate bugs, parasitic wasps, soldier beetles, spiders, syrphid flies.
Tidy Tips *Layia platyglossa*	**M** Aster. General companion plant. Adaptable, long-blooming, drought tolerant.	**BB, HB, SB, BI** Carpenter bees, mason bees visit this flower in its native California.	Assassin bugs, lacewings, lady-bugs, syrphid flies.
Toothpick Weed *Ammi visnaga*	**M** Carrot. Plant near chives, leek, lettuce, onion, rosemary, sage, tomato.	**BB, HB, SB** Small-tongued bees.	Big-eyed bugs, lacewings, ladybugs, minute pirate bugs, parasitic wasps, syrphid flies.
Verbena, Woolly, and Blue Vervain *Verbena* spp.	**M–L** Verbena. General companion plant. Many flowerheads on one stalk.	**BB, HB, SB** Long- and short-tongued bees, including cuckoo, mining, sweat bees.	Damsel bugs, lacewings, minute pirate bugs, parasitic wasps, spiders.
Vetch *Vicia* spp. Cultivars and native species.	**M–SUC** Legume. Fixes nitrogen.	**BB, HB, SB** Some are honey plants.	Ladybugs, minute pirate bugs, predatory and parasitic wasps.
Yarrow *Achillea millefolium* Cultivars and native species.	**M–L** Aster. Boosts other aromatic herbs, soil amender. Plant material speeds up compost.	**BB, HB, SB** Mining, sweat bees. Wool carder bees gather hairs along the stems to line nests.	Damsel bugs, lacewings, ladybugs, predatory wasps, syrphid flies.
Zinnia *Zinnia* spp.	**M–L** Aster. Japanese beetle trap crop. Avoid cauliflower, tomato.	**BB, HB, SB** Digger bees.	Hummingbirds, ladybugs, parasitic wasps, syrphid flies.

Although marigolds are beneficial to the garden, avoid double flowers lacking pollen and nectar. These waste bees' time and energy.

Bees of All Stripes

CHAPTER THREE

*T*he rainbow of colourful food on your plate is thanks to a bounty of bees in various colours, shapes and sizes. In fact, native bees and European honeybees play an integral role in the pollination of many plants, including at least a third of our fruits and vegetables. Yet a growing number of North American native bees are endangered or extinct. We need to protect those that are left—our future on this planet depends on it.

Most people can name about three different kinds of bees: honeybees, bumblebees and mason bees. Amazingly, there are 20,000 different kinds of bees, most of which don't produce honey, don't live in beehives and (thankfully) don't sting. Many wee bees are so small and gentle it is difficult to identify them. The smallest is *Perdita minima*, less than 0.08 inch (2 mm) long. The largest, Wallace's giant bee (*Megachile pluto*), is 20 times this size at 1.5 inches (3.8 cm) long and 2.5 inches (6.4 cm) wide.

Bees are divided into 9 families, 6 found in Canada and 7 in North America. There are 730 known species of bees in Canada alone, and 4,000 in North America. Once your eyes are opened to the rich biodiversity within the bee world, you may be able to discover at least 10 species right in your own backyard.

BECOME A BEE SPOTTER

To know bees is to love bees, and bee spotting is a useful skill for citizen scientists—amateur folks who help collect data on bees in their neighbourhood to share with professional scientists who map trends in pollinator populations.[22] While the taxonomy can be complicated and intimidating, as long as you can start to see the differences among bees and appreciate their

Opposite: A turquoise sweat bee (*Agapostemon*) forages on the flat centres of fleabane, an easy-access bee plant in the aster family.

A nest of Hunt's bumblebees (*Bombus huntii*) from research in blueberry fields by the Elle Lab at SFU.
MICHALINA HUNTER PHOTO

beauty, you will develop a growing passion for bees of all stripes.

Once you learn to identify the ones buzzing through your neighbourhood, you can find ways to support them; this is especially important for the species that require specific plant materials for food and nesting. For example, if you spot small carpenter bees in your neighbourhood, you can help them by planting native bushes in the rose family to provide stems for the bees to nest inside, nectar to feed adults and pollen for their offspring. Those rose bushes will support other species of bees in your garden. Leafcutters will cut circles out of the leaves to line their nests. Bumblebees and mining bees will "swim" in the shallow flowers, also collecting pollen for their brood.

A-TISKET, A-TASKET, HAS SHE GOT A POLLEN BASKET?

Some bees carry pollen in baskets, known as "corbiculae." These include honeybees, tropical stingless bees, orchid bees and bumblebees. A corbicula is a hollow depression on a bee's leg where the bee packs in pollen, which is held in place by special hairs. When fully stuffed, they look like saddlebags. Look closely at the hind legs of the bee. Chances are, if you spot corbiculae stuffed with pollen, she's either a true bumblebee or a honeybee. Most other bees collect pollen with their "scopa," which is a collection of special branched hairs (called "setae"). Many solitary bees are pollen-pants bees, with the scopa on their legs, and others are hairy-belly bees, with the scopa on the underside of their abdomen. Observe grooming behaviour and the colour of the pollen for insight into its foraging habits. Honeybees and bumblebees are generalists, but they also have pollen preferences. Other food preferences such as nectar and floral oils also give clues when identifying and categorizing bees. Males don't have structures for carrying pollen, and here's some good news: boy bees can't sting!

QUEENS, WORKERS, DRONES AND SINGLE MOMS

One of the ways bees are categorized is by their social structure. European honeybees are fully eusocial with a high level of social structure, while bumblebees are primitively eusocial with a lower level of social structure; the majority of bee species are solitary. While some solitary bees like to live together in groups, they can be very territorial about their own nest. Following mating, solitary females provision their own nests. Many solitary male bees throw slumber parties and sleep in or on flowers together.

HOLES, HIVES AND HOTELS

Bees are also categorized by their nesting behaviour. In rural and wild habitat, 30 percent (including honeybees, bumblebees and mason bees) are opportunistic cavity-nesters, finding and modifying existing chambers that suit their needs. The majority (more than 70 percent, including mining bees and sweat bees) are ground-nesting bees that excavate a hole or tunnel. In the urban environment, the percentages are flipped: 70 percent are cavity-nesters, with only 30 percent ground-dwelling, due to the lack of undisturbed sites for ground-nesting bees.

PHENOLOGY OF LIFE CYCLES: THE BEE CALENDAR

The majority of bees have an annual cycle, meaning the adults die within one year and leave eggs or queens behind to start the new generation. Annual bees can live from a very short cycle of a few weeks up to several months. Some emerge from their nests of wood, stems or soil early in the spring to mate, forage, lay eggs and die, while others emerge later in summer with life cycles that extend into September. European honeybees are perennial—meaning the hive continues as a community and does not die out—because they work collectively to store honey and pollen to allow them to live over the winter months or times of dearth. Even though the worker bees live less than a year, the queen can thrive for five to six years. Her sons and daughters come and go in a succession of generations that build up the hive structure and protect their fertile queen over the winter. She begins laying eggs in late winter so that there will be new adult bees ready to begin working when the spring blossoms open. In rare cases, bee species produce two generations in one year (bivoltine) or one generation every two years (semivoltine).

Bees are divided into seasonal guilds (groups of species that overlap in their use of environmental resources), depending on when they emerge and how long they live. Bees also have different diurnal behavioural patterns, foraging at different hours during the

Gourmands and Gluttons

There are technical terms for the different kinds of relationships bees have with pollen:

* Generalists are **polylectic**.
* Specialists are **oligolectic**.

Bee species fall somewhere on a spectrum from strict, narrow preferences for pollen to broad preferences for forage.

Oligolectic bees are further divided by just how specific they are:

* **STRICT OLIGOLEGE:** larvae need the pollen of preferred plants to develop properly.
* **WEAK OLIGOLEGE:** the bees gather pollen from their primary preference if available, but larvae can develop with other pollens.

day. Unlike most bees, bumblebees have the ability to thermoregulate (regulate their body temperature) and work early in the morning and later at night, while other species, such as squash bees, toil for limited hours in the morning or wait until the afternoon sun is at its peak.

WHEN PICKINGS ARE SLIM . . .

Most plants have evolved to attract more than one species of pollinator because a specialized relationship between plant and pollinator offers slimmer chances for the survival of both. However, a very limited floral preference, or oligolecty, is common in the Andrenidae (mining bee) family and the Halictidae (sweat bee) family. Plants that have a relationship with at least one species of oligolectic bee include cactus, *Ceanothus*,

A yellow-headed bumblebee with extended tongue flies over a California poppy. Wildflower meadows are important habitat for bumblebee conservation. JACK TUPPER PHOTO

goldenrod, maple, morning glory, nightshade, purple prairie clover, squash, sunflower, *Phacelia*, primrose, *Vaccinium* and willow. Making sure you include plants from these groups in your garden will help these vulnerable bees with a very selective palate.

FREQUENT-FLIER MILES

Bees vary widely in their maximum foraging distances from their nest sites. The smallest bees stick quite close to home, so they need an abundance of plants where they nest. The species that travel the farthest, such as honeybees and bumblebees, need pit stops along their routes, so they can refuel while gathering pollen.

- **BUMBLEBEES (*BOMBUS*):** 1.5 mi (2.4 km)
- **SMALL CARPENTER BEES (*CERATINA*):** 600 ft (180 m)
- **LARGE CARPENTER BEES (*XYLOCOPA*):** 1 mi (1.6 km)
- **EUROPEAN HONEYBEES (*APIS MELLIFERA*):** 1.9–5 mi (3–8 km)
- **LEAFCUTTER BEES (*MEGACHILE*):** 1,500 ft (450 m)
- **MASON BEES (*OSMIA*):** 300 ft (90 m)
- **MINING BEES (*ANDRENA*):** 1,500 ft (450 m)

- **PLASTERER BEES (*COLLETES*):** 500 ft (150 m)
- **SWEAT BEES (*HALICTUS*):** 600 ft (180 m)
- **SWEAT BEES (*LASIOGLOSSUM*):** 600 ft (180 m)
- **SMALL SWEAT BEES (*LASIOGLOSSUM [DIALICTUS]*):** up to 160 ft (50 m)

The tongue of a European honeybee is about 0.27 inches (7 mm) long. The bee with the longest recorded tongue is a tropical orchid bee called *Euglossa natesi*, with a tongue twice the length of its body at 1.32 inches (33.5 mm).

TONGUE LENGTHS AND NECTAR ROBBERS

The length of its tongue (proboscis) plays an important role in determining whether or not a bee can access nectar from a flower. Some nectars are so desirable that short-tongued bees will defy odds in accessing it by biting or poking a hole through the back of the flower,

A selection from T'ai Roulston's pinned bee collection, arranged from largest to smallest. T'AI H. ROULSTON PHOTO

insects, honeybees need even more space to store honey and pollen and provide space for over 50,000 sisters and brothers during peak season. Cavity-nesters seek their shelter in wood, clay, reeds, soil or other natural materials; some wait for other wildlife or humans to excavate tunnels and holes, while others industriously make their own. The majority of cavity-nesters find sites above ground, but bumblebees go below the surface to inhabit rodent nests. Bumblebees are eusocial ground-nesting insects, with each bee retaining some independence but living socially with the colony in a nest. The population of the nest (between 50 and 500 bees) is determined by the size of the nesting cavity and available forage.

BUMBLEBEES (*BOMBUS*): BUZZY BODIES
FAMILY Apidae · **SUBFAMILY** Apinae · **GENUS** *Bombus*

Yellow-faced bumblebees (*Bombus vosnesenskii*) have been implicated in the displacement of other bee species.

slipping in their tongue to slurp it up. The primary nectar robbers, bees that actually poke or chew holes in flowers, are large carpenter bees, tropical stingless bees and bumblebees, while other bees and insects are secondary robbers, taking the opportunity to steal nectar through existing holes.

Shorter-tongued guilds include mining, plasterer and sweat bees. Medium- to long-tongued guilds include bumblebee, large carpenter, long-horned, mining, honey, leafcutter, mason, oil-collecting and wool carder bees.

CAVITY-NESTING BEES— BOTH SOCIAL AND SOLITARY

Most cavity-nesting bees are solitary, requiring tunnels only slightly larger than their bodies to lay eggs and provision them with pollen. Honeybees and bumblebees, however, are social and need larger homes with rooms to feed, breed and tap dance. As highly social

Bumblebees are often heard before they are seen, especially inside the echo chamber of a bell-shaped flower or bloom that requires the special services of buzz pollination. There are several species of bumblebees that are declining, endangered or extinct, which is why it is critical to meet the needs of these bees.

With 250 known bumblebee species globally and nearly 50 in North America, these hairy bees are generally quite pleasingly plump and 0.4 to 0.9 inches (10 to 23 mm) in length. Some of the little ones in the brood may be runts that received the smallest share of pollen in the nest.

Bumblebees come in many colours and patterns. Some key clues to identifying their species are variations of abdominal stripes and thoracic markings in yellow, orange, black and white on their back ends. The males are shaggier than females and in some species have a much lighter colouring. If you are searching for males, look for blonds with no pollen baskets. These bees can't sting, so if you're willing to trust your identification skills, you can impress your friends by capturing a male bee in your hand. You may see females mix the pollen with their saliva as they groom it into their corbiculae.

Bumblebees are opportunistic nesters and will defend their nests collectively, but otherwise are rarely provoked to sting. (You are most likely to be stung by a bumblebee when walking barefoot on a lawn that contains clover.) Once a bumblebee colony reaches a certain population density—ranging from 50 to 500 bees—additional queens develop and mate before hibernation in the fall. The new queens of summer are the last to stop foraging and burrow into mulch to hibernate.

Only the new queens survive the winter; the rest of the colony dies. Hibernating bumblebee queens are among the first native bees to emerge in the spring to begin foraging and establish a nest. When the queen emerges at the beginning of the cycle in early spring (or later, depending on the species), she goes on a hunt for the perfect nest location: a dry, sheltered place that contains insulating materials. Abandoned rodent nests are a common choice, and queens are also known to

The Wandering Life of the Male Bumblebee

Once male bumblebees reach maturity, they leave the nest to begin their life's work: sipping nectar, mating with young queens and finding a sweet spot to spend the night. Males set up a territory, which they map with the tantalizing aromas of their pheromones, marking prominent objects in the landscape such as boulders, leaves, fences and tree trunks. They patrol their territory and renew the pheromone markers in the morning and after it rains. In order to prevent confusion and congestion on their highways of love, some species leave their scent markers as low as 3 feet (1 m) from the ground, with others patrolling higher up in the trees. While honeybee drones die after mating with a queen, bumblebee males can mate several times, pouncing on an unmated fertile female to release sperm and then excreting a fluid that hardens into a plug that prevents other males from gaining access to her.

nest in rock cavities, compost, attics, the insulation between walls, birdhouses, old couches, and even under lawn mowers. Happily, they also occasionally deign to use the nesting boxes put out by optimistic bee scientists.

The queen makes her brood nest herself, secreting flakes of wax from her abdomen. In the spring (or later, depending on the species) she sits on a nest of about 10 of her eggs and warms them up—she can raise her internal temperature up to 95F (35C) in the same manner as a broody hen. After she's raised a critical number of brood that can go out to help provision the nest, the queen stays inside to lay eggs that her daughters help care for. Bumblebees are progressive feeders, giving the larvae bits of food at a time.

Generalist foragers, they are divided into guilds depending on tongue length and the timing of their life cycle, with each guild needing a constant source of appropriate plants.

How You Can Help Bumblebees

- Create sheltered piles of rotting leaves for queen bumblebees to dig down into and hibernate in the fall until they emerge in early spring.
- For long-tongued bumblebees, plant flowers with long nectar tubes from the borage, figwort, heather, honeysuckle, legume, mint, plantain and teasel families.
- Medium-tongued bumblebees are fond of plants in the amaryllis, rose, mint, willow and aster families.
- For bumblebees of all stripes, plant anise hyssop, boneset, campanula, chives, crimson clover, comfrey, currants, goldenrod, honeysuckle, hyssop, Joe-Pye weed, larkspur, lavender, liatris, lungwort, monarda, penstemon, prairie clover, purple toadflax, ratibida, red clover, rudbeckia, perennial sage, pincushion flower, self-heal, snapdragons, snowberry, vetch, Virginia bluebells and wild delphiniums.
- Plant flowers that require buzz pollination, such as those in the heather and nightshade families.

Mistaken Identa-Bee

Be on the lookout for all sorts of bumblebee wannabees out there, including bee flies, robber flies, clearwing moths, cuckoo bumblebees and giant carpenter bees.

Male bumblebees can often be found having a nap on a leaf or a flower. Sometimes they wrap their legs around a plant part and lock in place with their mandibles, but other times they nestle into the centre of a flower where the temperature can be up to 50F (10C) warmer. Morning glories and other nyctinastic flowers close up on sleepy bees in the evening and release them when sunlight returns. Once the flower refills with nectar, the bee can drink breakfast and be on his way.

LARGE CARPENTER BEES (*XYLOCOPA*): SAWDUST MAKERS

FAMILY Apidae · **SUBFAMILY** Xylocopinae · **GENUS** *Xylocopa*

A gentle non-stinging male valley carpenter bee (*Xylocopa varipuncta*). The metallic black females look strikingly different.

Unlike the bees that use prefab homes, carpenter bees are industrious DIY-ers that dig out tunnels to make nests. Females drill the holes in soft wood to keep their family safe. Males guard the home with fly-by warnings to potential predators . . . but since they lack stingers, their buzz is worse than their bite. Females have a sting but it is much milder than that of bumblebees. The two genera earning the name "carpenter"—large and small carpenter bees—vary widely in size. The large carpenter bee, at least as big as a bumblebee queen, is like a flying teddy bear with a dense coat of fur even woollier than that of the bumblebee.

There are 400 species of *Xylocopa* globally, with 32 in North America and only 1 in Canada—the eastern carpenter bee (*Xylocopa virginica*), found in southern Ontario. Hairy and stout with dark wings, they are 0.5 to 1.2 inches (13 to 30 mm) in length, some with blue eyes and others green. The females have large mandibles for chewing wood to make nests, mixing sawdust with spit to build walls between nest chambers. They can be dark brown, black, metallic blue or metallic green, and carry dry pollen on the scopae (a mass of specialized hairs) on each of their hind legs. Males are sometimes dark golden with large eyes and an ability to hover, which is part of their charming lekking behaviour, resembling the old courting ritual of a box social. Males gather in groups around food sources and each bee claims a plant to defend and mark with rose-scented sexual pheromone. Each female chooses the perfumed perch of her choice and waits for the male to mate with her; if he's not hovering about, he misses his chance.

While large carpenter bees are mostly solitary, some are semi-social. Cavity-nesters, they find shelter in the

> ### *Lekking for Love*
> "Lekking" comes from the word "play" in Swedish, but in Dutch and Danish it can also mean "luscious," with overt sexual connotations. Some species of male bees hover and perch around their territory, jostling with other males for prime positions and permitting the female bees to feast their eyes on their magnificent physiques. It's hard work being eye candy, which is why when you see restless bees posing rather than provisioning, you'll know they're likely male bees lekking for love.

stems of yucca and agave in the south, along with conifers and sheds and garden furniture made from untreated wood. Reliable tunnels are sometimes re-used from year to year. New bees overwinter as adults, and females can live up to three years. Large carpenter bees are long-tongued generalists and nectar robbers, choosing large flowers. Depending on the species, they can be found from spring to fall and are important pollinators of many food crops, including tomatoes and other nightshade plants that they buzz-pollinate. Tragically, some folks consider these DIY bees a nuisance and respond with deadly pesticides. If there is a strong concern about where carpenter bees have settled, a humane removal expert can be called in to relocate them instead.

How You Can Help Large Carpenter Bees
~ In places conducive to long-term nesting, leave out pieces of deadwood, particularly cedar, pine

Opposite: Flowers in the aster family provide landing and refuelling pads for bees. **Clockwise from top left:** Chocolate cosmos, false sunflower, sneezeweed and elecampane.

and spruce. Drill deep holes 0.5 to 0.7 inches (13 to 18 mm) in diameter.

- Plant flowers in the rose and legume families, as well as aster, bidens, blueberry, blue-eyed grass, boneset, brown-eyed Susan, cactus, catalpa, germander, goldenrod, gooseberry, highbush huckleberry, honeysuckle, Joe-Pye weed, lavender, liatris, lupins, milkweed, mints, monarda, nightshade, passion flower, penstemon, phacelia, purple coneflower, ratibida, sage, self-heal, snowberry, sumac, sneezeweed, Solomon's seal, sunflower, swamp dogwood, thistle, tomato, wild delphinium, willow and wisteria.

Mistaken Identa-Bee
Large carpenter bees can be confused with bumblebees.

SMALL CARPENTER BEES (*CERATINA*): ROSE HOBOS

FAMILY Apidae · **SUBFAMILY** Xylocopinae · **GENUS** *Ceratina*

Small carpenter bees are the right size to sip nectar from the tiny florets in an aster blossom.

If you see a bee poking her head out from inside a dead raspberry cane, it just might be a small carpenter bee. There are 351 species of *Ceratina* in the world, 21 in Canada and the US. These are small to medium bees, 0.1 to 0.6 inches (3 to 15 mm) in length with cylindrical abdomens. They are dark metallic blue-black in colour, often with white, yellow or ivory face markings. Females carry pollen on scopae on their hind legs.

Small carpenter bees are mostly solitary, nesting inside stems of plants, often in the rose family. Cell divisions are made of saliva and plant pulp scraped from the walls of nests. A female uses her body to plug the nest hole before she dies, leaving it in place over the winter to protect her offspring as they overwinter as adults. Small carpenter bees have medium to long tongues and can be found foraging in spring and summer in small flowers. One of the most common bees in eastern North America is *C. calcarata*, a small carpenter bee that nests in sumac twigs and raspberry canes.

How You Can Help Small Carpenter Bees
- Plant sunflowers and shrubs in the sumac and rose families for stems.
- Lightly prune dead stems in your roses and raspberries for nests, revealing the pithy interiors.
- Plant flowers in the aster and carrot families to bloom through spring and summer.
- For forage, plant buttercup, green-headed coneflower, potentilla, spring blue-eyed Mary, strawberry, willow, verbena and violets.

Large carpenter bees lay the largest eggs in the insect world: 0.6 inches (15 mm) long.

Mistaken Identa-Bee
Small carpenter bees are easily confused with small sweat bees.

EUROPEAN HONEYBEES (*APIS MELLIFERA*): A-LIST CELEBRITIES

FAMILY Apidae · **SUBFAMILY** Apinae ·
SPECIES: *Apis mellifera*

Some honeybees collect pollen as they forage for nectar; others purposefully scrape it off the anthers of the flowers.

There are 10 to 12 species of honeybees in the world, none native to North America. While there are different kinds of honeybees in Africa and Asia, it is the European honeybee that is farmed for pollination and honey in North America and Europe. Africanized "killer" honeybees are a result of cross-breeding the African honeybee (*A. mellifera scutellata*) with European honeybees.

European honeybees are medium-sized hairy bees 0.4 to 0.6 inches (10 to 15 mm) in length. They are amber to dark brown (almost black) with stripes on their abdomens, some quite pronounced depending on genetic lineage. In fact, when you hear beekeepers talking about "the Italians" or "the Germans," chances are they are comparing bee genetics rather than football statistics. You will notice that some honeybees are translucent and retain an amber glow in the sunshine. Healthy bees have a fuzzy appearance, with the newborns as fluffy as chicks. Male bees (drones) are bigger than females, with larger eyes to track down the queen for mating. While female worker bees carry pollen in the corbiculae on their hind legs, the pollen collects on the branched hairs over their entire bodies, even their eyes.

Cavity-nesters in wood and custom-built hives, European honeybees are highly social, with colonies of thousands of bees comprising a fertile queen bee, infertile worker bees and drones. Female workers have many jobs throughout their lives, including raising the young bees and creating stored food in the form of honey and "bee bread." Also known as "ambrosia," bee bread is made of pollen mixed with nectar. The queen lays eggs in hexagonal honeycomb cells made of wax extruded by workers. Active from spring to fall and year-round in warmer climates, honeybees are generalist foragers with medium to long tongues.

In farmers' markets in Cambodia, honeybee larvae are sold alongside honey for human consumption.

How You Can Help Honeybees
- Wherever possible, plant one or two acres of organic honeybee pasture for every hive.
- Let Dutch clover, crimson clover and dandelions flourish.
- Plant trees such as poplars, which honeybees use

to collect sap for propolis, a substance used to seal and protect hives.

Mistaken Identa-Bee

Honeybees are confused with wasps, squash bees, drone flies and syrphid flies.

LEAFCUTTER BEES (*MEGACHILE*): JELLY BELLIES

FAMILY Megachilidae · **SUBFAMILY** Megachilinae · **GENUS** *Megachile*

A large leafcutter bee performs synchronized swimming exercises in the lush purple blossoms of a thistle.

Leafcutter bees have sharp mandibles to cut leaves and petals into small circles that are glued together and sealed with bee saliva and sometimes resin to create partitions for their nest tunnels. They are opportunistic nesters, using existing cavities in wood, rock, stems and even snail shells. Canada has native New World leafcutter bees and imported Old World alfalfa leafcutter bees (*M. rotundata*) that are managed to pollinate livestock forage on the Prairies. If you see a bee carrying a piece of leaf or petal as she's flying, you know it's a leafcutter making her nest. You may find some leafcutter cocoons made of pieces of rolled leaves in your mason-bee condos. Just leave them be and they should hatch out next summer.

Of the more than 1,500 species of *Megachile* leafcutters globally, 140 are native to North America. Leafcutter bees are small to large: 0.4 to 0.8 inches (10 to 20 mm) with wide heads to accommodate prominent jaws. Distinctive, with narrow pale stripes tapering to a point, their abdomens are flattened and sometimes tilt up at the back end. Males of some species have long hairs on their forelegs, which they place over the female's eyes while mating. Females have white or yellow scopae on the underside of their bellies and carry dry pollen.

Leafcutter bees often lift their "booty" in the air when collecting pollen. There should be a yoga pose devoted to this gesture—"upward bee" as opposed to "downward dog." They are methodical foragers, circling around from floret to floret sipping nectar. The bees in the Megachilidae family, which includes leafcutter and mason bees, have been nicknamed "jelly belly" bees because of the bright patches of pollen on their bellies.

Leafcutter bees are solitary, occasionally nesting in aggregates. Opportunistic cavity-nesters, they utilize crevices and tunnels made by other insects in soil, rotting wood, hollow stems, under bark, and in any manmade structures that mimic their natural homes, such as the hollow spaces between cedar shingles and holes drilled into logs.

Appearing in summer and fall, some leafcutters are generalists while others are specialists, preferring the aster and legume families. Look for wings that are spread apart while foraging, a distinctive habit with these bees. Some of the most important agricultural pollinators are in this family, pollinating blueberries, carrots and onions, and in the case of the alfalfa leafcutter bee (*M. rotundata*), the alfalfa fed to livestock.

How You Can Help Leafcutter bees

- ~ Leave deadwood in your yard and drill holes in logs.
- ~ Grow plants leafcutters use to line their nests, such as roses and other broadleaf deciduous species.
- ~ Plant allium, anise hyssop, aster, boneset, campanula, coreopsis, goldenrod, Joe-Pye weed, liatris, milkweed, penstemon, purple coneflower, red clover, rudbeckia, sea holly and sunflower.

MASON BEES (*OSMIA*): BUILDER BABES

FAMILY Megachilidae · **SUBFAMILY** Megachilinae · **GENUS** *Osmia*

You can provide nesting condos in your garden for blue orchard mason bees. MARK MACDONALD PHOTO

Mason bees are so named because they make walls of clay between eggs in their nest tunnels. Very difficult to photograph, they rarely sit still: males are usually searching for females to mate with and females are busily gathering pollen and mud and laying eggs, only stopping when they are jumped on by a randy male. The females are super-pollinators, outperforming honeybees in the orchards where they are put to work by

Bees on the Balcony

Beekeeper Brian Campbell tells the story of when his neighbour came by with a potted plant—a small rose with little hole-punch–sized dots cut out of many of the leaves. Just then, a little leafcutter bee stuck its head out and seemed to say, "I didn't know this was a *mobile* home." Brian gave the little guy a secret salute, said, "It doesn't seem to be hurting the plant, so I would just leave it alone," and sent his neighbour on his way.

farmers. You'll only need one small mason bee condo to ensure the pollination of the trees in your yard, and many neighbouring trees as well. So, while you won't reap honey from keeping mason bees, they will reward you with healthy fruit galore.

There are 350 known species of mason bees in the genus *Osmia* worldwide and 135 in North America. Mason bees are small to medium, 0.3 to 0.6 inches (8 to 15 mm) in length, stout and rounded and often very dark with a blue or blue-green metallic sheen. Some have wide heads to accommodate the prominent jaws they use to collect mud. They are hairy-belly bees with scopae on the underside of their abdomens to carry dry pollen. Males are smaller and thinner than the females and have a white goatee. Females have darker hair on their faces and a mild sting. Unlike drones (male honeybees), male mason bees feed themselves. The adult females drink nectar for energy and store extra nectar in a honey sac to mix with pollen to feed to their larvae.

Mason bees are solitary, sometimes nesting in aggregates. Most nests are found naturally in crevices and tunnels made by other insects in rotting wood, under bark, in hollow stems, and in any man-made structures that mimic their natural homes, such as the hollow spaces between cedar shingles and mason bee condos. Most mason bee condos available commercially are for the blue orchard mason bee. The bees make partitions out of mud, sometimes mixed with chewed plant materials.

There are spring and summer guilds of mason bees. Some of the most important agricultural pollinators are in this family because they are fast and efficient foragers. There are two main species of blue orchard mason bees in Canada, *O. lignaria propinqua*, native to British Columbia, the Prairies and Arizona, and *O. lignaria lignaria*, native to provinces east of Manitoba and to Georgia. Specialist mason bees include *O. ribifloris*, also known as the blueberry bee, and *O. collinsiae*, a specialist pollinator of spring blue-eyed Mary (*Collinsia verna*). Other mason bees include *Hoplitis*, specialists in the legume, mint and figwort families, and *Chelostoma*, specialists in the waterleaf family (a subfamily of the borage family).

How You Can Help Mason Bees

- Provide mud that's rich in clay.
- Leave deadwood in your yard.
- Install and maintain a mason-bee condo.
- Grow plants in the barberry, borage, figwort, legume, mint, rose, carrot and willow families.
- For *Osmia* mason bees, plant apple, bluebells, blueberry, *Ceanothus*, cherry, *Collinsia*, dandelion, Dutch clover, dwarf larkspur, fleabane, forget-me-not, hardy geranium, honeysuckle, laburnum, lupin, lyreleaf sage, overwintered brassica, quince, peach, pear, penstemon, *Pieris japonica*, plum, *Sisyrinchium*, snowberry, strawberry, thistle, verbena, violet, Virginia bluebells, wild delphinium and wild garlic.

bee byte

The mason bee *Osmia conjuncta* makes nests in snail shells.

Mistaken Identa-Bee
Mason bees are often mistaken for houseflies.

RESIN BEES (*HERIADES*): PINE NESTERS
FAMILY Megachilidae · **SUBFAMILY** Megachilinae · **GENUS** *Heriades*

A British resin bee (*Heriades truncorum*) carries sticky plant material to seal her nesting tunnel. JEREMY EARLY PHOTO

A few of the smallest hairy-bellied bees have a special talent for making the cell partitions of their nests with the resin from evergreen trees, some with a special predilection for pine. There are 140 species of these small resin bees worldwide, 11 in North America. Delicate and thin, these small to medium bees are 0.16 to 0.28 inches (4 to 7 mm) long. Resin bees are dark grey to black, with thin white hair bands on their abdomens.

Some females have orange scopae on their abdomens, which often curls under at the tip.

Small resin bees are cavity-nesting, mostly in wood, some in stems. They use resins, sometimes mixed with sand, tiny pebbles or chewed plant material, to make partitions. These short-tongued bees are summer-foraging generalists. They are important pollinators of apple, bean, coffee, cowpea, eggplant, tomato, sunflowers and more. *H. carinata* is common in eastern Canada. Resin bees (*Heriades* and *Dianthidium*) are also sometimes classified as mason bees.

How You Can Help Resin Bees

- Set out pine stumps, drilling them with small holes in various sizes. (You could do this with your used Christmas tree!)
- Grow pine trees, plants in the aster family, bidens, catmint, *Ceanothus*, coreopsis, fleabane, goldenrod, milkweed, prairie clover, rudbeckia, smooth hydrangea, sumac, sunflower and thistle.

Heriades occidentalis sometimes nests in pine cones.

Mistaken Identa-Bee
Easily confused with small leafcutter and mason bees.

WOOL CARDER BEES (*ANTHIDIUM*): WOOLLY BULLIES

FAMILY Megachilidae · **SUBFAMILY** Megachilinae · **GENUS** *Anthidium*

Large male wool carder bees mate with smaller females as they are foraging.

There are 170 known species of *Anthidium* globally, 5 in Canada and 36 in the US. They are stout bees, 0.3 to 0.8 inches (7 to 20 mm) long, with males considerably larger than females. Wool carder bees have distinct yellow or white markings on black or dark-grey abdomens. The females have light-coloured hairy bellies. Some species have green eyes. A wool carder

Wasps are often mistaken for bees, but yellow jacket (*Vespula* spp.) larvae are carnivorous and bees are vegetarians.

bee can be picked out by its hovering flight pattern and aggressive habit of mid-air hip checks. The clear black and yellow markings make them easy to confuse with wasps if you see them at a glance. Males hang out on leaves and wait to defend their territory. They will bump other bees right out of flowers, even pinning down rivals with the ridged tips of their abdomen.

Wool carders are fast-flying generalists but, as long-tongued bees, they are big fans of plants in the mint family for nectar. They are solitary cavity-nesters in wood and stems, and some species dig their own nests in sandy soil. Female bees use plant hairs from woolly lamb's ear, mullein and *Phacelia bolanderi* to line nests and form walls between chambers. Unlike other tunnel-nesters, male eggs are laid at the back of the tunnel, with the female eggs in front. Each cell is packed with pollen the female gathers on her belly, and the nest is sealed with bits of debris and chewed plant pulp. You may find wool carder fluff and pupae in your mason bee condos. The species native to North America usually forage in spring, while the European wool carder (*Anthidium manicatum*) is active in summer, when it tends to move into your garden and monopolize plants with fuzzy leaves.

How You Can Help Native Wool Carder Bees
~ Wool carder bees from Europe have been introduced into North America, competing directly with native bees. If you live in an area with native North American wool carder bees (such as *Anthidium palliventre*), you can support them by growing spring-blooming plants in the mint and pea families, along with plants with woolly foliage. Otherwise, it would be best not to encourage an overabundance of European wool carder bees.

Mistaken Identa-Bee
Carder bees are easily confused with wasps and *Dianthidium* bees, which are closely related to resin bees.

GROUND-NESTING BEES—SWEET AND MOSTLY SOLITARY

While roughly 70 percent of bees nest in the ground, each species has a particular preference for the kind of soil they choose and the angle of the land, from flat to steeply sloped. These sites should be protected from human activity and grazing whenever possible. These bees have hairy legs to gather up pollen and, although most are solitary, some like to nest in groups. Sweat bees, one group of ground-nesters, are among the most common bees in North America and include some of the most photogenic—the green sweat bees that dazzle in the bright sunlight with an iridescent sheen. Most sweat bees are sweet, delicate little solitary insects that play an integral role in the pollination of a wide variety of plants.

MINING BEES (*ANDRENA*): ROSE SWIMMERS

FAMILY Andrenidae · **SUBFAMILY** Andreninae · **GENUS** *Andrena*

A large bumblebee forages next to a mining bee (*Andrena*) in an Edmonton backyard. HC PROCTOR PHOTO

Andrena mining bees are very common in the garden, foraging alongside honeybees and mason bees in overwintered brassicas. When the wild roses bloom in May and June, they swim in the stamens, collecting pollen on their hairy legs. There are 1,500 known species in the world, with about 450 in North America, north of Mexico. Consisting of small to medium bees 0.3 to 0.7 inches (7 to 18 mm) in length, *Andrena* is a very diverse genus, found in a variety of colours including black, dull metallic blue, red or green, with yellow to silvery white hairs and often pale bands of hair on their abdomens. The smallest species are difficult to identify in the field. Females have velvety eyebrows and carry pollen on scopae high on their upper hind legs, making it appear as if they are up to their armpits in pollen. Their weak stingers cannot pierce your skin.

Andrena are solitary, sometimes forming aggregates in favourable nesting sites, and a few species nest communally with females sharing a nest. Most species line the soil with a waxy secretion. Depending on species, they nest in flat or gently sloped bare ground or patchy lawns in sandy or silty soils. A few species raise more than one generation in a year. Most are generalist foragers from spring to fall, with spring guilds being more abundant. They are important pollinators of food crops such as blueberry, cranberry, sunflower and onion. Some species are more cold-hardy than honeybees, making them important northern pollinators of early spring crops such as cherry and apple.

How You Can Help Andrena *Mining Bees*

- Create patches of flat and sloped sandy, silty soil and leave bare patches in the lawn.
- Grow a variety of plants rich in pollen and nectar from spring to late fall in the aster, carrot, rose and willow families, particularly goldenrod.
- Plant flowers they pollinate, including allium, apple, blueberry, bog rosemary, cherry, cranberry, hawthorn, milkweed, snowberry, spiraea and sunflower.

Mistaken Identa-Bee

Andrena mining bees can be mistaken for honeybees and *Halictus* sweat bees.

Mining bees (*Andrena*) perform together like synchronized swimmers, bathing in the generous pollen of wild roses.

DIGGER BEES (*ANTHOPHORA*): TINY TERRIERS

FAMILY Apidae · **SUBFAMILY** Apinae · **GENUS** *Anthophora*

This long-tongued digger bee (*Anthophora terminalis*) was napping in my mom's garden in Cactus Lake, SK.

There are about 460 known species of *Anthophora* bees globally, 70 species in Canada and the US. These digger bees are hairy and stout, some with grey/silver hairs and black stripes on their cylindrical abdomens. They range greatly in length, from 0.25 to 1 inch (6 to 25 mm). Some have blue or green eyes. Males are smaller and thinner than females, with lighter-coloured faces that often have yellow markings. They sometimes sleep in aggregates and emerge before females to patrol the nesting ground in groups. They help the females dig themselves out, only to pounce on them to mate. Females carry pollen on scopae on their lower hind legs.

Anthophora are solitary ground-nesters with the one exception of *A. furcata*, which nests in rotting wood. They dig out nests like enthusiastic tiny terriers kicking soil out behind them. Nests are dug in flat ground and hard clay, on sandy beaches or in banks of loam or sandy loam, and are lined with a waxy water-resistant secretion. Offspring are fed the pollen and nectar *soup du jour*. Pupae do not make cocoons, and some bees overwinter as adults. *Anthophora* digger bees are fast-flying generalist foragers appearing from March to June. With medium to long tongues, they are important pollinators of cherries and tomatoes.

How You Can Help Digger Bees

- Leave bare soil and deadwood for nesting sites.
- Plant members of the aster family for pollen and members of the nightshade, borage and mint families for nectar, particularly lavender, catnip and perennial sage.
- *Anthophora* are found on bidens, bugloss, cherries, coreopsis, cosmos, currant, Dutch clover, Dutchman's breeches, fleabane, germander, hardy geranium, hoary puccoon, honeysuckle, milkweed, nightshades, penstemon, phacelia, Russian sage, shooting star, Solomon's seal, spring blue-eyed Mary, vetch, violet, and Virginia bluebells.

One species, *Anthophora abrupta,* is called the "chimney bee" because it creates an external tunnel at the entrance to its nest.

Mistaken Identa-Bee

Anthophora are easily confused with long-horned bees.

LONG-HORNED BEES (*MELISSODES*): MINI JACKALOPES

FAMILY Apidae · **SUBFAMILY** Apinae ·
GENUS *Melissodes*

A furry long-horned bee (*Melissodes*) poses for the paparazzi on blanket flower (*Gaillardia*). SEAN MCCANN PHOTO

These long-horned bees are the cuddly jackalopes (mythical horned rabbits) of the mining bees. Exclusively New World bees, with 130 known species and 120 in North America, they are stout, furry and small to medium, 0.3 to 0.7 inches (8 to 18 mm) long. Some species have noticeable pale stripes, others are velvety ash-grey and black, and some have striking blue or blue-green eyes. Males have long antennae, and sometimes sport yellow brushy moustaches. They hang out in aster and daisy flowers looking for females. Females are larger than males with wider bodies and carry pollen on scopae on their hind legs. *Eucera* and *Svastra* are also long-horned bees.

Solitary and occasionally nesting in aggregates, these ground-nesters line their tunnels with waxy secretions. Medium- to long-tongued and often aster or daisy specialists, they are important pollinators of wild and cultivated sunflowers in late summer and fall. They also like to sip nectar from lavender, catnip and other plants in the mint family.

How You Can Help Long-Horned Bees

- Leave bare soil in your yard.
- Grow flowers in the aster and daisy families that bloom in late summer and fall: aster, bidens, blanket flower, boneset, brown-eyed Susan, Canada germander, cosmos, cup plant, goldenrod, ratibida, tickseed, and wild and cultivated sunflowers.
- Grow plants in the mint family for a nectar source.
- The long-horned sunflower bee (*Svastra obliqua expurgata*) prefers ratibida and sunflowers.

Mistaken Identa-Bee

Sunflower bees (*Svastra* spp.) may be mistaken for *Melissodes*.

An elegant silver cultivar of globe thistle (*Echinops ritro*) blooms at the University of British Columbia Botanical Garden.

PLASTERER BEES (*COLLETES*): POLYESTER PANTS

FAMILY Colletidae · **SUBFAMILY** Colletinae ·
GENUS *Colletes*

A dew-covered aggregate of sleeping male plasterer bees (*Colletes*). SEAN MCCANN PHOTO

There are 469 identified species of *Colletes* in the world, but it is suspected that there are at least 200 additional ones. The 107 species of plasterer (a.k.a. polyester or cellophane) bees in Canada and the US range in length from 0.3 to 0.6 inches (8 to 15 mm). They are furry little creatures with convex striped abdomens and adorable heart-shaped faces. Males are thinner and longer than females, with white facial hair. The abdomen of females tapers to a point. Males sleeping in aggregates are the most loveable minifauna of them all.

Plasterer bees are solitary, occasionally nesting in aggregates. Females secrete linalool into the cells, which acts as a fungicide and bactericide. Since the nests are lined with this cellophane-like secretion, these bees can nest in sandbanks and creek beds, creating a series of brood chambers. The plastic is secreted from the Dufour's gland at the rear of the bee, next to the gland that creates poison for her stinger. Brood provisions are a soupy mix of pollen and nectar. Generations of bees tend to return to sites over many years. Females are polyester pollen-pants bees, with scopae on the upper hind legs and thorax, giving them the appearance of being up to their armpits in pollen. With short two-lobed tongues, plasterer bees can be generalists or specialists with preferences in the aster family or nodding onion. Spring guilds forage in maple and late summer guilds luxuriate in goldenrod. The goldenrod cellophane bee (*C. solidaginis*), a rusty, shaggy teddy bear, is a goldenrod specialist.

How You Can Help Plasterer Bees

- Leave bare soil in your yard.
- Plant goldenrod, maple, nodding onion and plants in the aster, borage, mallow, legume and willow families that bloom from spring to fall.
- Plant herbs containing linalool, such as basil, cilantro, lavender and monarda.

bee byte

Linalool is used in many soaps and personal hygiene products, as well as by pest-control companies to repel fleas, fruit flies and cockroaches. It is also found in some fungi.

Mistaken Identa-Bee

Colletes plasterer bees are easily confused with *Andrena* mining bees.

MASKED BEES (*HYLAEUS*): MYSTERIOUS MINERS

FAMILY Colletidae · **SUBFAMILY** Hylaeinnae · **GENUS** *Hylaeus*

In this photo of mating *Hylaeus*, you can see the more pronounced mask on the male bee. HC PROCTOR PHOTO

Masked bees are a type of polyester or plasterer bee. There are over 700 known species of masked bees globally, with 51 in North America, north of Mexico. Red alert: more than two dozen species of masked bees are currently on the Xerces Society's Red List of Bees (a system rating declining bee populations from fragile to extinct) due to the loss of their native habitat.

Small, slender and 0.2 to 0.3 inches (5 to 7 mm) in length, masked bees are shiny and black with yellow or white markings on the face and legs; the masks are more noticeable on the male bees. What sets masked bees apart from other bees is that they are hairless because they carry pollen internally in a crop. This means that the provisions they regurgitate for their brood are quite moist compared to the pollen balls and bee bread that other bees make for baby food.

Most species of masked bees are solitary cavity-nesters in stems and galls, deadwood and the abandoned nests of other insects. They commonly nest in raspberry canes, which they line with cellophane-like secretions, and a few are ground-nesting. Masked bees are short-tongued generalist foragers, with some specialists that forage on plants in the rose family. They generally forage in small flowers, but can lap at the full nectaries of larger flowers. Most species are active in spring, with some persisting into fall.

How You Can Help Masked Bees
- Prune raspberry canes and roses, placing the prunings in your insect hotel.
- Add plants in the aster, carrot, rose and willow families, and let them bloom.
- Plant the flowers masked bees are mostly found foraging on, such as anise hyssop, blue-eyed grass, boneset, brown-eyed Susan, *Ceanothus*, fleabane, foxglove, goldenrod, hawthorn, long-headed coneflower, lupin, milkweed, mock orange, mountain mint, penstemon, phacelia, potentilla, prairie clover, sage, strawberry, sumac, virgin's bower clematis, willow, wild garlic and yarrow.

Mistaken Identa-Bee
Masked bees are easily confused with small wasps and small carpenter bees, but generally are less shiny.

The sculptural blossoms and seed pods of *Nigella hispanica* are strikingly beautiful.

OIL-COLLECTING BEES (*MACROPIS*): LOOSESTRIFE LOVER

FAMILY Melittidae · **SUBFAMILY** Melittinae · **GENUS** *Macropis*

An oil-collecting bee (*Macropis nuda*) forages on fringed loosestrife (*Lysimachia ciliata*). JOEL GARDNER PHOTO

There are some bees with extremely fine-tuned relationships to one family of plant, or even one species of flower. These oligolectic (specialist) insects are as vulnerable as the wildflowers they evolved alongside over millennia. Such is the case with the *Macropis* bees, forage specialists in the loosestrife (*Lysimachia*) genus within the primrose family. Tragically, humans are destroying the habitat of these bees. There are only about 16 species worldwide and 2 in eastern Canada (*M. ciliata*, and *M. nuda*). A species in the US, *M. steironema opaca*, is likely extinct. Even rarer than the oil-collecting bee is the *Macropis* cuckoo bee (*Epeoloides pilosulus*), endangered in Canada. Oil-collecting bees are small to medium and 0.3 to 0.6 inches (7 to 15 mm)

long. They are dark brown-black with silvery hairs. Males have yellow face markings and longer antennae than females, and females have pollen-collecting hairs on their hind legs and extra-long hairs for collecting floral oils. These hairs extend farther down the leg than those of other bees.

Oil-collecting bees are solitary, but in rare exceptions females may share nests. Nesting sites are usually found in the sandy soil adjacent to wet, swampy areas that host loosestrife. Long-tongued foraging specialists, they are dependent on the flowers of *Lysimachia* for larval provisions. Females use the floral oils to line cell walls and mix them with pollen to provision the cells. Adults sip nectar from other flowers for energy. They are summer bees, appearing when the flowers they depend on are blooming.

How You Can Help Oil-Collecting Bees

- Leave bare soil in your yard.
- Plant wildflowers in the primrose family. These include the loosestrifes that are native to eastern North America, not to be confused with the Old World invasive purple loosestrife. The New World loosestrifes include fringed loosestrife (*Lysimachia ciliata*), swamp candles (a.k.a. yellow loosestrife/*L. terrestris*), tufted loosestrife (*L. thyrsiflora*), and more. Fringed loosestrife is of particular importance if you live in the native area of this bee (Ontario and Nova Scotia).
- Plant milkweed, which also supports monarch butterflies.

Mistaken Identa-Bee

Andrena mining bees can appear to be look-alikes of oil-collecting bees.

HOARY SQUASH BEES (*PEPONAPIS*): NECTAR NOMADS

FAMILY Apidae · **SUBFAMILY** Apinae · **GENUS** *Peponapis*

Male squash bees sleep in the blossoms of pumpkins. SCOTT SMITH PHOTO

Squash bees are rare in Canada, with only one species found in Ontario and Quebec, hoary squash bees (*Peponapis pruinosa*). This species spread to Canada as Indigenous societies brought squash cultivation with them from the south. There are 13 species in North and South America. Hoary squash bees are medium-sized, 0.5 inches (13 mm) in length, with golden-brown hair and darker brown stripes. Males have long antennae and yellow markings on the lower face. Females are larger than males, with darker faces. They are pollen-pants bees, with brown setae forming the scopae on their hind legs.

Hoary squash bees are solitary, nesting in the flat ground around squash plants. They are summer bees, named for their foraging preference, and females forage earlier in the morning than honeybees. Squash bees are an example of a bee that has evolved with one kind of plant, and the life cycle of these bees is in synchronicity with that of the squash. The plant is a bed and breakfast for the males, which sleep in groups in the flowers at night, feed on the pollen and nectar and mate with females as soon as the flowers open in the wee hours of the morning. All this has happened before the honeybees have even ventured outside their hives. These bees are endearing, with their fuzzy thoraxes and furry chaps on hind legs that accumulate the large pollen grains from squash. Other squash bees include *Xenoglossa,* which has evolved to fly in the dark before sunrise. These bees are so efficient at pollinating cucurbits that they can work together with other native pollinators like bumblebees to pollinate fields of squash without help from honeybees. The decline of squash bees has been linked to pesticides.

How You Can Help Squash Bees
- Leave flat, bare soil in your yard.
- Plant a variety of cucurbits in your organic garden: cucumbers, gourds, pumpkins, squash and watermelons.
- Grow buttonbush (*Cephalanthus occidentalis*), milkweed, morning glory and verbena.

Mistaken Identa-Bee
Squash bees are easily confused with honeybees, long-horned bees and sunflower bees.

Blanket flower, calendula, cosmos and sneezeweed blossoms float in a bee oasis at City Farmer in Vancouver.

GREEN SWEAT BEES (*AGAPOSTEMON*): GLAM ROCKERS

FAMILY Halictidae · **SUBFAMILY** Halictinae ·

GENUS *Agapostemon*

A photogenic green sweat bee (*Agapostemon*) forages in five spot.

There are 43 identified species of green sweat bees in North America. They are small to medium bees 0.3 to 0.6 inches (7 to 15 mm) in length. In spite of being grouped with the sweat bees, these species do not actually lick your sweat. The bees can be metallic green or blue, some with black and white or black and yellow stripes on their abdomens. The males (and some females) of a few species have the jaunty stripes on their abdomens, while some females are one solid iridescent colour. Females carry pollen on scopae on their hind legs.

Green sweat bees can be solitary or nest communally. The females make nests in soil, preferring sandy loam, sometimes on slopes. One nest tunnel may serve up to 24 females, each making her own set of chambers off to the sides of the main entrance. Green sweat bees are fast-flying generalists with short tongues, foraging for pollen and nectar in a wide variety of flowers. The name *Agapostemon* means "stamen lover." You will find some green sweat bees that emerge in late spring foraging in campanulas and others in the fall foraging in asters. Some species have more than one generation in a year. *A. texanus* has a larger range than other species in this genus.

How You Can Help Sweat Bees

- ⤳ Leave areas of bare soil and protect nesting sites from livestock. Plant shallow open-access flowers in the aster, carrot, mint and rose families.
- ⤳ Green sweat bees are found on aster, blazing star, boneset, Canada germander, catnip, compass plant, coreopsis, cup plant, Dutch clover, false sunflower, foxglove penstemon, goldenrod, grey dogwood, hardy geranium, Joe-Pye weed, long-headed coneflower, mallow, morning glory, mountain mint, prairie clover, purple coneflower, rudbeckia, snowberry, sunflower, smooth sumac, sneezeweed, thistle, verbena, wild hyacinth (*Camassia scilloides*), wild monarda and yarrow.

Mistaken Identa-Bee

Agapostemon can be easily confused with another kind of shiny green sweat bee, *Augochlora*, which has a rounder and smoother abdomen.

Glittering jewel wasps like this one can also be mistaken for green sweat bees.

SWEAT BEES (*HALICTUS*): TICKLE LICKERS

FAMILY Halictidae · **SUBFAMILY** Halictinae ·
GENUS *Halictus*

A *Halictus* sweat bee forages in a pink potentilla blossom. HC
PROCTOR PHOTO

Halictus sweat bees are true sweat bees, seeking salt from human and animal sweat. There are currently 200 identified species worldwide, with 10 species in North America. They are delicate, thin and small to medium in size, at 0.2 to 0.6 inches (5 to 15 mm) long. *Halictus* sweat bees are dark brown to black, many with a metallic sheen. The hairy stripes are thinner in the centre of the abdomen, thicker at the sides. Males have longer antennae and sometimes yellow markings on the face and legs. Females carry pollen on scopae on the upper part of their hind legs.

    Halictus sweat bees are solitary or semi-social. Some species have more than one generation per year, with the daughters remaining in the nest to care for their younger sisters. The bees nest in flat ground, preferring sandy and silty soil, lining the nest cells with a waxy excretion from their Dufour's gland that makes the cells waterproof. When it comes to foraging, sweat bees are generalists, and these mighty little bees are important pollinators of crops like sunflowers and watermelon. Orange-legged furrow bees (*H. rubicundus*) are common throughout the northern hemisphere. This bee is solitary or semi-social, depending on the climate of the location where it makes its nest.

How You Can Help Halictus *Sweat Bees*

∿ Grow plants in the aster, carrot, mint and rose families as well as flowers they are found in, such as catmint, clover, dogwood shrubs, mountain mint, persicaria, prairie clover, sunflower, verbena and watermelon.

Mistaken Identa-Bee

Halictus sweat bees are easily confused with *Lasioglossum*.

SWEAT BEES (*LASIOGLOSSUM*): PINT-SIZED POWERHOUSES

FAMILY Halictidae · **SUBFAMILY** Halictinae ·
GENUS *Lasioglossum*

A *Lasioglossum* sweat bee forages in calendula.

If there are 1,700 identified species of *Lasioglossum* in the world, and 280 in North America, how come folks don't know more about these wee bees? They deserve to have their own pollinator week, or at least a moniker that celebrates their pollinating prowess. The smallest species are difficult to identify in the field, ranging in length from 0.1 to 0.4 inches (3 to 10 mm). *Lasioglossum* are black, brown, green or blue, often with a dull sheen. In species that have stripes, the bands of hair are thicker in the centre of abdominal segments. Males are thinner than females, with longer antennae and yellow markings on the lower part of their faces. Sometimes the boys sleep in aggregates. Females are larger than males, with darker faces. These are pollen-pants bees, carrying pollen on scopae on their upper hind legs.

It is not surprising, with so many species, that these tiny bees have variations in social structure, including solitary, semi-social and social nesting preferences. *Lasioglossum* nest in soil or (rarely) in rotting wood. Females line and mark the entrance of the nest with waxy lactones (chemical compounds found in certain plants). Although most *Lasioglossum* are foraging generalists, there are specialists favouring primrose and clarkia. These bees visit many agricultural crops, including strawberry and other fruits in the rose family, sunflower and other crops in the aster family, asparagus, sweet potato, cucurbits, and okra and other plants in the mallow family. Species can be found foraging from spring to fall. *L. oenotherae* is a specialist pollinator of common and yellow sundrops. It's found in eastern Canada from Ontario to New Brunswick and in some bordering states.

How You Can Help Lasioglossum *Sweat Bees*
- Leave bare soil and rotting wood for nests and protect nesting sites from livestock.
- Plant a variety of plants with smaller flowers from spring to fall, including evening primrose and clarkia species.

Mistaken Identa-Bee
Lasioglossum are easily confused with *Halictus* sweat bees.

CUCKOO BEES—BEE IMPERSONATORS

A cute cuckoo bee (*Coelioxys*), easily identified by its sharply pointed abdomen. SEAN MCCANN PHOTO

A cuckoo bee is a bee mimic—a solitary predator that lays her eggs in a bee's nest and then buzzes off to drink nectar. Some species are very good bee impersonators; others look more like wasps because they lack scopae. You will often find them hovering around the nest entrances of the bees that they parasitize—they sneak into the nest and lay eggs that hatch and grow into larvae that eat the provisions left for the original eggs. The cuckoo bees grow into adults in the host nest and emerge to drink nectar from the same plants as the bees they mimic. They do not collect pollen, though, because they have outsourced their childcare and have no need of it.

Cuckoo bees and predatory crab spiders like this one are a sign of healthy biodiversity in the bee garden.

CUCKOO BUMBLEBEES (*PSITHYRUS*): TREASON WHISPERERS

FAMILY Apidae · **SUBFAMILY** Apinae · **GENUS** *Psithyrus*

There are 29 species of *Psithyrus* (Latin for "murmuring"). About the same size as the queen bees they usurp, female cuckoo bumblebees can attack the colony they are invading, killing the queen or subduing her with a chemical cocktail of pheromones. The cuckoo bumblebee destroys the eggs and larvae of the deposed queen and recruits the worker bumblebees in the hive to raise her own offspring. Many are excellent mimics, but can be distinguished by shiny "bald spots" and pointy abdomens. You will see these solitary predators sipping nectar alongside true bumblebees and hovering around their nest entrances in abandoned rodent burrows or other similarly cozy cavities.

Do We Really Want to Help Cuckoo Bees?

~ Yes! The more bee plants you include in your garden, the more fascinating visitors will drop by for lunch and stay for dinner. And even though cuckoo bees prey on true bees, hosting them in your garden is a sign of healthy biodiversity. Best left alone, they help to keep balance in the garden and are part of the natural environmental pressures that actually help create genetically strong survivor bees.

Mistaken Identa-Bee

Obviously, in some cases it is easy to mistake a cuckoo bee for the true bee it mimics; in other cases, cuckoo bees can be mistaken for wasps or flying ants because they lack the scopae for collecting pollen that make many bees distinctive.

CUCKOO LEAFCUTTER BEES (*COELIOXYS*): COCOON PIERCERS

FAMILY Megachilidae · **SUBFAMILY** Megachilinae · **GENUS** *Coelioxys*

There are about 500 species of cuckoo leafcutter bees, a.k.a. sharp-tailed bees, in the world. Leafcutter cuckoo bees are medium to large and 0.3 to 0.6 inches (7 to 15 mm) in length. Male cuckoo leafcutter bees look very similar to leafcutter bees, but may have sharp spines on the end of their abdomens. Females have wide heads and a black and grey conical abdomen ending in a point used to pierce the cells of leafcutter bees to lay eggs. When the larvae hatch, they eat the eggs of the true bee. These solitary predators can be found foraging in the same flowers as true leafcutters or hovering around nest entrances in tunnels in deadwood and hollow stems.

CUCKOO SWEAT BEES (*SPHECODES*): HOME INVADERS

FAMILY Halictidae • **SUBFAMILY** Halictinae • **GENUS** *Sphecodes*

There are 300 species of cuckoo sweat bees globally, 80 in North America. They are small, slender, black and 0.2 to 0.6 inches (4 to 15 mm) in length. Some can be identified by a red abdomen, and males have white facial hair. Their exoskeleton is designed to act as armour to protect them from the attacks of their hosts. *Sphecodes* are active late spring to early fall, laying eggs in the nests of sweat bees. The offspring of these solitary kleptoparasites kill the host larvae and eat their food. They can be found foraging in the same flowers as true sweat bees and hovering around tiny nest entrance holes in bare ground and scrappy lawns.

RED CUCKOO BEES (*NOMADA*): RUBY ROBBERS

FAMILY Apidae • **SUBFAMILY** Nomadinae • **GENUS** *Nomada*

There are 700 species of red cuckoo bees worldwide and 287 in North America. Small to medium-sized, from 0.2 to 0.5 inches (5 to 15 mm), with slender bodies, they appear very similar to wasps. Some red cuckoo bees have ivory, yellow or white markings and thick antennae. In some species, males mimic *Andrena* hormone to lure female *Nomada* bees to nest sites, presumably to mate. The cuckoo bee then sneaks in to lay her eggs, where her larvae kill their host's larvae. Red cuckoo bees appear in spring, parasitizing sweat and mining bees (mostly *Andrena*). They can be found foraging in the same flowers as mining and sweat bees, hovering around the tiny nest entrance holes in bare ground and patchy grass.

If it looks like someone has taken a hole punch to your rose leaves, chances are you've got leafcutter bees in your garden.

MELISSODES CUCKOO BEE (*TRIEPEOLUS*): ARMOURED STEALTH

FAMILY Apidae · **SUBFAMILY** Nomadinae · **GENUS** *Triepeolus*

❊

There are 150 species of *Triepeolus* globally, 108 in North America. They are medium-sized bees, 0.3 to 0.7 inches (7 to 18 mm) in length, with distinctive black and grey markings. Their hairless armoured bodies make them look ready to battle. Parasitizing many species of bees, most often *Melissodes*, they can be found sipping nectar neck-and-neck with long-horned bees and hovering close to the ground waiting to slip into the nests of these ground-nesters to lay eggs.

BEE MIMICS—THE WANNABEES

Like cuckoo bees, there are other insects that mimic bees and are part of the beautiful biodiversity that keeps your garden in balance. It is thought that bee mimics protect themselves from predators by Batesian mimicry, which means appearing to be more dangerous than they really are. There are three quick ways to tell the difference between flies and bees: flies have two wings while bees have four; flies have short, stubby antennae, while honeybees have long, elbowed antennae; and flies have large compound eyes near the front of the head, while bees have two compound eyes at the sides of the head and three small compound eyes on the forehead.

BEE FLIES (*BOMBYLIIDAE*): FLYING FLUFFBALLS

Very fuzzy with two large composite eyes, bee flies are 0.04 to 1 inch (1 to 25 mm) in length and range from dark rusty-brown to grey, black or blonde, with a few species sporting jaunty moustaches. Some bee flies are so hairy that it is difficult to see a division between the thorax and abdomen. They have long tusk-like tongues, earning them the nickname "beewhals," and feast on nectar and pollen. The larvae parasitize other insects in their nests, and adults can be found hovering around the entrance to mining bee nests in spring and summer.

ROBBER FLIES (*ASILIDAE*): MOUSTACHIOED ASSASSINS

This carnivorous robber fly looks so much like its prey that it's hard to tell them apart. SEAN MCCANN PHOTO

There are 7,000 known species of robber flies (a.k.a. assassin flies) worldwide, with more than 1,000 in North America. They have large compound eyes and fuzzy moustaches. Ranging from 0.4 to 2 inches (10 to 50 mm), these insectivores are diverse in appearance, with the more delicate species attacking prey smaller than that of the larger robust furry killers. Adults ambush their prey, piercing them with their mouthparts, injecting a paralytic toxin and then sucking out the insect's juices. In addition to bees, robber flies kill aphids, Mexican bean beetles, Colorado potato beetles, Japanese beetles and grasshoppers.

SYRPHID FLIES (*SYRPHIDAE*): APHID ANNIHILATORS

Syrphid flies are bee mimics that come in many shapes and sizes.

There are about 6,000 species of syrphid flies (a.k.a. flower flies or hoverflies), with almost 900 in North America. They can be small to large, ranging from 0.16 to 1 inch (4 to 25 mm) in length. As wasp and bee mimics, they are often yellow, amber or brown, with black markings, and can be quite elegant. Some syrphid flies have a similar amber colour to honeybees, which bewitches and befuddles the bee watcher, but syrphid flies hover over flowers rather than making a beeline from one to another. Most syrphid flies are less hairy than honeybees, which makes them glow even more brightly in the sun.

Adults feed on nectar and pollen; larvae have varying diets, depending on the species. Some feed on decaying plant matter, while others are insectivores, preying on aphids, caterpillars, leafhoppers and thrips. Like bees, some species are more adapted to deep nectar tubes while others prefer shallow flowers.

PLANTING ✽ PLAN ✽

A Rooftop Garden for Bees

Hardy native plants exploding with colour create a pollinator paradise that will turn any drab space into a vital work of art. Rooftop gardens with indigenous blooms are a way of rewilding the city and boosting populations of urban native bees. These gardens help give inner city children their first glimpse of wildflowers that are becoming a rare sight, even in the countryside. As our cities become densely populated, we've got to plan for pollinators by creating a network of roof-to-roof oases for bees. Set up the lawn chair and enjoy the fireworks!

SNOWBERRY CLEARWING MOTH (*HEMARIS DIFFINIS*): MASTERS OF DISGUISE

Another bumblebee mimic is a large day-flying moth, which is found in the Northwest Territories, British Columbia and southern Ontario, along with much of the US. Its furry body is gold and black and 1.25 to 2 inches (32 to 50 mm) in length. Its wings are clear with striking black markings. It is also known for resembling other creatures, earning it the names "hummingbird moth" and "lobster moth." Larvae feed on cherry, hawthorn, honeysuckle, mint, plum, snowberry and viburnum. Caterpillars pupate inside cocoons in piles of decaying leaves. Adults sip nectar from honeysuckle, lilac, snowberry, thistles and violets.

Healthy Herbs for Pollinators and People

CHAPTER FOUR

*T*he Latin names for many herbs contain the word *officinalis,* which comes from *officina,* referring to the room where herbs were traditionally stored in a monastery. Throughout history, herbs have had a vital connection to bees, and monks were among the many who tended herbal gardens to help their bees thrive. In fact, honeybees were kept so close to herb and vegetable gardens that they were frequently sheltered in special recessed niches called "bee boles," created in the walls of manors and cottages to house straw hives known as "skeps." While many of the Old World herbs we use in cooking today are still closely associated with honeybees and the cloistered gardens of days gone by, bumblebees and native bees of all stripes also benefit immensely from these aromatic and nectar-rich plant apothecaries.

Opposite: Anise hyssop and goldenrod make a dynamic duo to feed bees in late summer.

HERBS AS WHOLE FOODS

Growing a beauteous and healthful bounty of herbs for edible flowers, foliage and seed will save you trips to the grocery store, reduce your carbon food-print and give bees a giant boost.

Warning: All herbs should be consumed in moderation. In particular, pregnant women and those with

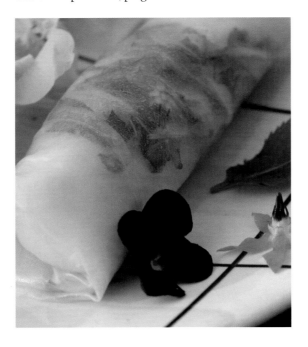

PLANTING PLAN

The Victory Hub: A Bee Garden for the Senses

Gardens lure us in with the pleasure they give to our senses. Luckily for us, bees are also drawn to the scent, shape, colour and smell of the flowers that fill humans with delight. Herb patches on the grounds of schools, seniors' homes and community centres have a stimulating and healing function for body and soul. Tasting and touching aromatics like sage and mint can improve memory, mood and concentration. One very meaningful benefit of historical Victory Gardens was the morale-boosting effect of working on the plots—in times of stress and darkness, the scent of honeysuckle and herbs and the comforting buzz of bees can help to restore a sense of peace and equilibrium. See a sample Victory Hub bee garden plan on opposite page.

specific health concerns are advised to check with their physician prior to consuming herbs. Please refer to the planting chart at the end of this chapter for some specific contraindications.

PETALS TO PLATES

While most recipes calling for herbs suggest leaves, flowers are just as tasty and provide a gorgeous garnish too. And placing an emphasis on using flowers in

My husband makes salad rolls with edible fresh flowers such as nasturtium, violet, borage, calendula and scarlet runner bean.

the kitchen ensures herbs will bloom for the bees. If you have herb flowers aplenty, many—anise hyssop, bee balm, calendula, dragonhead, lavender, mallow and more—can be dried and tossed into tea blends to add depth of flavour.

LEAVES THAT KEEP ON GIVING

One of the great gifts of herbs is that, once you harvest a few sprigs for your kitchen or medicine cabinet, the plant will grow back prolifically, usually bushier than before. These "cut-and-come-again" herbs are a good garden investment, especially if you have limited space.

SEEDY HERBS FOR BEES

Growing herbs for seed ensures bees have a chance to consume optimum pollen and nectar rewards from flowers. The seeds of many herbs—anise, fennel, lovage, nigella, onion and poppy, to name a few—are tasty in teas and sweet or savoury dishes. Excess seeds can also be grown as flavoursome and healthy microgreens or tossed in blank areas of the garden for bee-supportive bursts of bloom.

MEDICINE-CABINET HERBS FOR BEES—ESSENTIALS

British herbalist and beekeeper Juliette de Baïracli Levy believed all beekeepers should plant a garden of herbs to support their hives. In *The Complete Herbal Handbook for Farm and Stable* she notes that bees are skilled at treating themselves with herbs, and benefit from healing properties in the nectar and pollen they ingest and harvest from particular plants. Herbal

A black-tailed bumblebee (*Bombus melanopygus*) rests on a blue nigella blossom. Nigella flowers and seedpods are highly decorative and self-seeding.

beekeepers agree, and use herbal teas and essential oils to benefit hive health. De Baïracli Levy also noted that honey made from herbs has special healing properties for both bees and humans.

Both honeybees and wild bees benefit from the chemical properties of herbs and the copious nectar and pollen offered by their flowers. De Baïracli Levy observed that bees particularly love aromatics. She stressed the importance of including mints, especially peppermint, in the beekeeping garden. Umbels are invaluable too, she says, such as herb fennel with its alluring licorice scent. And nectar-rich plants are vital to a healing bee garden, with a special emphasis on the deep blue blossoms of borage (*Borago officinalis*). Generations of Mediterranean beekeepers have relied on their sun-loving herbs to support honeybee health, particularly such lemony powerhouses as lemon balm and dragonhead, along with aromatic classics such as lavender, oregano, rosemary, sage and thyme.

AWESOME UMBELS FOR BEES

Umbels are nature's ultra-efficient way of packing many flowers into a small space, saving foraging bees time and energy. Some also form perfect landing pads, making it easier for pollinators to park and sip.

Angelica

An Old World umbel, Angelica belongs in any herb garden for bees. My favourite cultivar of angelica is "Vicar's Mead," with purple stems and aromatic white flowers. Adding stems to rhubarb crumble adds depth to the dish and enough sweetness that you can cut back on the sugar. Bees are sweet on angelica too, its small blossoms attracting small- and medium-tongued nectar sippers.

Anise

A natural bee lure, anise looks and tastes similar to fennel, and its tiny seeds can be used in South Asian curries and chai spice mixes. Anise is a traditional beekeeping herb; a few drops of anise oil mixed with water in a spray bottle helps calm bees and can be used instead of smoke. (It's also good for helping the beekeeper stay calm.) Some beekeepers add anise oil to the sugar syrup they use to feed bees in times of dearth. Since anise is so attractive to bees, it can also be part of a mixture used to condition a swarm trap or new hive. Rub anise oil or plants inside the clean hive or outside the entrance.

Cilantro

When cilantro is grown for its tasty seeds instead of the delicious greens, it is referred to as coriander. However you identify this plant, its tiny, delicate white florets are ideal for tiny bees with short tongues. Sowing this

8 feet (2.4 m) before blooming in midsummer, place it where it will not block the light for the rest of your plants. Checking first that it is not providing habitat for developing ladybugs, give the stems a stubble cut in the fall, leaving 8 to 12 inches (20 to 30 cm) behind for cavity-nesting bees; bundle the remainders of the stems into bee hotels (see Create Hollow-stem Hotels on page 108).

herb every two weeks from spring through early summer ensures a continuous bloom for bees and a bounty of leaves and seeds for your own culinary creations. As a bonus, varroa mites are repelled by the essential oil from the seeds.

Fennel

Fennel is allelopathic, producing chemicals that inhibit the growth of other plants, so provide it with a private stage in a corner of the garden where it will flourish and attract pollinators galore. Trim off the licorice-flavoured seeds for use in cooking and as a tasty digestive tea— but look before you snip. If you are lucky, your fennel plant may have become a nursery for developing ladybugs and should be left alone until they fly away home.

Lovage

A great companion plant for your garden, lovage improves the flavour of most vegetables that grow around it and attracts ground beetles and ladybugs that prey on microfauna that munch your prized plants. Lovage looks and tastes like celery on steroids and adds depth to soups and stews. Since it grows up to

Toothpick Weed

One of the essential umbels for bees and beneficial insects, toothpick weed gets its name from one use for the stems. It is also sold as "honeyplant" or "visnaga." Seeds are used to treat stomach ailments, and leaves and blossoms can be added to salads. A row of the lacey flowers will act as a busy insectary and bee highway, making it more possible for smaller insects and bees to float around your garden.

LEMON AID FOR BEES

Lemony-scented herbs help bees send signals to one another, making it easier to negotiate the complex world of colour and aroma they need to navigate to mate, forage and bring home the pollen.

Lemon Balm

While lemon balm can be too prolific a self-seeder at times, there's no denying its benefits for bees. Lemon balm contains geraniol, which attracts bees. In fact, bees use geraniol, among other aromatics gathered from plants, to mark nectar-bearing flowers and their hives. In the evenings, honeybees perch at the hive

entrance and, to call their sisters home, raise their abdomens and fan their wings to release a pheromone from the nasonov gland, which contains geraniol. Beekeepers have also been known to rub lemon balm on the entrances to beehives to help bees find their way back and to attract swarms. Geraniol is also very good for humans, since it is antioxidant and antibacterial and helps fight tooth decay.

In addition to dragonhead and lemon balm, beekeepers rub anise, fennel, hyssop and thyme on hives to welcome bees home.

Lemon-scented Bee Balm

Highly alluring to many bees and invaluable in the kitchen, bee balm (a.k.a. bergamot) is a New World plant with a similar scent to the Old World citrus bergamot, used to flavour Earl Grey tea. Lemon bee balm (*Monarda citriodora*) is an especially good match to the flavour of citrus bergamot. Use it to make Victory Garden "lemonade," along with lemon catnip, lemon thyme and lemon verbena.

Dragonhead

If you like lemon balm, you will love dragonhead, an Old World herb that makes a lovely lemon-scented tea. The flowers look like purple dragons' heads, which bees enter to collect nectar. Its geraniol content gives it a scent very similar to lemon balm, but unlike lemon

balm, it dries and stores well. It is also used by beekeepers to rub on hives as a homing scent. Dragonhead blooms over several weeks, making it a good gap filler for bees, especially if grown in a significant patch. Happily, it does not share lemon balm's tendency to voraciously self-seed.

MINT MAGIC FOR BEES

It's hard to beat mints for nectar production, blossom density and bee-attracting scents. Clearly this plant family has evolved a mutually beneficial relationship to its pollinating bees.

Scarlet and Wild Bee Balm

Scarlet bee balm (*Monarda didyma*) is a classic buzz-hummer, attractive to long-tongued bumblebee queens and hummingbirds. Wild bee balm (*Monarda fistulosa*) has lavender flowers with nectar tubes that are not as long as those of scarlet bee balm, providing nectar for a wider spectrum of bees.

Catnip and Catmint

Loved by felines, catnip is also one of the best flowers for bees. True catnip (*Nepeta cataria*) can be a prolific self-seeder, but if you have a tolerance for spreading plants, by all means don't fence it in. Other species of *Nepeta* (such as *Nepeta × faassennii*) are called "cat*mint.*" They are well-behaved perennials and some of the best plants for bees because they bloom continuously from spring to fall. Cut the plant down after flowering and it will bloom again. Catmint is a good source of nectar and pollen and will fill in gaps when not much else is flowering.

Mints

The group of plants called *Mentha* is spectacular for bees. Peppermint and spearmint are gateway mints, leading gardeners into a world of biodiverse scents and tastes such as chocolate, pineapple and ginger mint.

Opposite top: The lacey flowers of toothpick weed create a superhighway for bees and beneficial insects. **Opposite bottom:** Edible flowers and herbs make a platter of vegetarian sushi into a tasty work of art.

Plant a variety of mints to achieve a long period of bloom. As some varieties can take over the garden, you can grow mint in a pot, preferably raised above the ground. Or let it run rampant in an underused area of the yard. Beekeepers place mint sprigs inside hives to fend off ants and use peppermint oil to discourage varroa mites from moving in.

Some beekeepers have had luck rubbing mint lotion on their hands to keep from being stung as they work in the hives.

NECTAR-RICH NOURISHMENT FOR BEES

For those bees with large families to feed and long distances to travel, nectar-rich plants are a matter of survival. But even the smallest solitary bees need these life-giving liquids.

Since honey is evaporated nectar, it generally takes 110 pounds (50 kg) of nectar to produce 44 pounds (20 kg) of honey.

Anise Hyssop

One of the most drought-tolerant mints native to North America, anise hyssop is a New World plant that honeybees and bumblebees love—in fact, it is quite possibly the most favoured forage of honeybees and *the* top honey plant. Bees of all stripes perform gymnastics

> **ANISE HYSSOP HERBAL INFUSION**
>
> For a soothing beverage, add **2–3 tsp (10–15 mL) fresh anise hyssop leaves and flowers** (or half as much if dried) to **1 cup (250 mL) boiling water**. Steep for 10 minutes, strain and serve hot or cold. Try mixing anise hyssop with yarrow, rosehips, monarda, cinnamon, vanilla, rosemary, lavender or marjoram.
>
> ◁ **recipes** ▷

around the stamens to collect the precious pollen, and bumblebees grasp onto the ends of long flowers and use their weight to tip the nectar from the long corolla. Anise hyssop is easy to grow, doesn't spread like other mints, and can bloom from late June or July until the first frost when planting is staggered.

Borage

A classic edible and medicinal Old World bee plant, borage is easy to sow and fast to grow. Honeybees and bumblebees sip its nectar and collect ivory pollen in their corbiculae. Once established, the plants will self-seed and pop up throughout the season in the garden, providing a continuous supply of nectar and pollen summer through fall. The blossoms pump out nectar all day long and the heads hang down in the rain to protect the nectar from being diluted. Bumblebees buzz-pollinate the blooms and cover themselves in showers of blue-grey or off-white pollen. The flowers are edible, with a slight hint of cucumber flavour.

Hyssop

Not to be confused with anise hyssop, hyssop (*Hyssopus officinalis*) is an Old World nectar-rich plant with wide appeal to many bees. Traditionally used to flavour fatty meats and oily fish, the leaves have a much stronger

The busiest part of an herb garden is often the oregano patch, where honeybees can be seen foraging alongside native bees, like this long-horned bee (*Melissodes*).

What mint and other herb flowers lack in showiness they make up for in scent and flavour for bees.

taste than the flowers, and can be bitter or camphorous, so use in small amounts. Hyssop is one of the herbs in metheglin (spiced mead), along with rosemary, thyme, oregano and sage.

Mignonette

An Old World herb traditionally included in gardens near beehives, mignonette deserves to be revived in contemporary gardens. What the flowers lack in showiness is made up for by the sweet scent that drives honeybees wild, drawing them into your garden.

MEDITERRANEAN MEDICINE FOR BEES

Since bees are nectar- and solar-powered, it makes sense to nourish them with time-honoured bee plants from the sun-drenched hills of Europe.

Lavender

The ultimate bee magnet, lavender has such an alluring fragrance that it will draw bees into your garden,

and there is the added bonus that its oil can be used to repel varroa mites. A powerful moth repellent, lavender is an excellent orchard companion.

Oregano

One of the most concentrated nectars found in the garden, oregano's super medicinal essential oils and long bloom make it one of the best plants for bees of all stripes. Its active ingredients include thymol, which has aromatic and antiseptic properties, suggesting it could play a role in supporting bee health, repelling hive pests and guarding against diseases. If you have a sunny, well-drained spot in your garden with ample room for a plant to spread by self-seeding, tuck in some wild oregano and your garden will be humming with a mixed chorus of bees for weeks in the summer heat.

Rosemary

Beekeepers use rosemary and rose-scented geraniums to rub inside clean hives to make them attractive to

honeybees, while deterring wax moth. Rosemary contains camphor, a natural moth repellent, while the geraniums entice the bees with geraniol and nerol.

Sage (Culinary)

One of the most important essential plants in any herb garden for bees, sage contains antibiotic and antifungal oleic acid. Beekeepers add sage to the smoker to help calm bees, and brew a sage and chamomile tea as a spring tonic for bees. The purple blossoms provide medicinal nectar for long-tongued bees.

Thyme

There are a mind-boggling number of varieties of thyme, and since some varieties are small and compact, you can plug it into the empty spaces in any sunny, well-drained patches in your borders, lawn, rock walls or containers. Thyme is a powerful insect repellent and beekeepers use thymol, one of its active ingredients, in varroa mite treatments.

Opposite, clockwise from top left: From the pioneering years of beekeeping in North America, anise hyssop is a nectar-rich plant worth reviving. Blooming from May to the first frost, borage is an essential bee forage plant. A wool carder bee clasps a fragrant lavender stalk to sleep through the night. Hyssop (*Hyssopus officinalis*) was traditionally grown in monasteries and used medicinally on humans and honeybees. **Below:** Infusing herbs in honey creates a synergy of the healing properties of the herbs with honey's antibacterial powers. When infusing fresh herbs, store honey in the refrigerator and keep only for short periods.

EDIBLE HERBS
for Bees

A SEASONAL PLANTING CHART

*B*erbs and their edible flowers bestow benefits on both humans and bees. Just be sure to leave a good percentage to bloom to feed the bees and other beneficial insects. Practise selective bolting: allow a few cut-and-grow-again herbs to flower while snipping other leaves for your kitchen creations. Avoid double-flowered varieties of plants because they waste the bee's valuable energy without providing rich desserts. Some herbs are honey plants (plants that provide a large quantity of nectar suitable for honey) and vary widely in potential honey yield, depending on factors such as the weather and soil quality.

Warning: Herbs should be ingested with caution. This chart indicates usage of small amounts of edible plants for food only, not for medical purposes. It does not recommend ingesting essential oils, which can be toxic.

When planting from scratch, check the seed package for how deep to plant the seed or follow the rule of thumb: plant a seed at a depth roughly double its diameter. As noted, some seeds need light for germination. Heights for plants are maximum and vary according to weather and soil conditions; blooms/bolting have been taken into consideration, as leaving some plants to bloom for the bees is essential to your Victory Garden for Bees.

PLANT ORIGIN	BLOOM PERIOD	HARDINESS	BEES ATTRACTED
OW: Old World (indigenous to Europe, Asia or Africa)	E: Early	refers to the coldest zone the plant is hardy to.	BB: Bumblebees
	M: Mid-season		HB: Honeybees
	L: Late	DEER AND DROUGHT	SB: Solitary bees
NW: New World (indigenous to the Americas and their islands)	SUC: May be seeded in succession	RESISTANCE are most reliable once the plant is established.	BI: Beneficial insects

PLANT NAME	PLANT FAMILY, BLOOM, FOLIAGE	HARDINESS	HEIGHT AND GARDEN NOTES	BENEFITS TO BEES*	PERKS FOR PEOPLE
Angelica *Angelica archangelica* and *A. sylvestris* **OW**	**M** Carrot. Large globe-shaped umbels of dainty white to white-green flowers. Wild angelica (*A. sylvestris*) is invasive in eastern provinces.	**Biennial/ Perennial** Hardy to zone 4.	**3–8 ft (0.9–2.4 m)** Dappled shade in loamy or sandy, slightly acidic, well-drained soil. Moist conditions. Start from root cuttings, second-year offshoots or fresh seeds (need light to germinate). Taproot.	**BB, HB, SB, BI** Nectar. Tiny flowers attract mining bees, *Andrena* spp., and other small bees.	Edible hollow stalks taste like celery. Add flowers to salads and omelettes. Stems candied for an after-dinner digestive treat. *Avoid if pregnant or breastfeeding.*
Anise *Pimpinella anisum* **OW**	**M–L** Carrot. Tiny white flowers on dense umbels. Licorice scent.	**Annual**	**1–4 ft (0.3–1.2 m)** Light, fertile soil with good drainage, sunny location. Start seeds indoors in early spring. Long taproot. Water until established.	**HB, SB, BI** The scent of anise is particularly attractive to honeybees.	Leaves and seeds lend a licorice flavour to sweet and savoury dishes. Anise tea helps relaxation. *Avoid if pregnant or breastfeeding, or if you have cancer or are receiving hormonal therapies.*
Anise Hyssop *Agastache foeniculum* **NW**	**M–L–SUC** Mint. Small bilabial mauve flowers packed onto spikes. Licorice scent.	**Perennial** Hardy to zone 3. Deer and drought resistant.	**2–4 ft (0.6–1.2 m)** Sun to part shade in dry to medium loam with sand or rocky material. Seeds require moist stratification and light.	**BB, HB, SB, BI** Honey potential: 350 lbs/ac or much more. Digger, leafcutter, mining, masked bees.	Leaves and flowers used fresh or dried to make soothing teas with a licorice flavour; added to fruit salad.
Basil, Holy, a.k.a. Tulsi *Ocimum sanctum* **OW**	**M** Mint. Light-purple flowers. Aromatic foliage is green or purple.	**Tender Annual**	**2–3 ft (60–90 cm)** See sweet basil. Originally from India, the strain called "Kapoor Tulsi" is recommended for North American climate.	**BB, HB, SB** Significant medicinal honeybee plant.	Holy basil tea made from leaves is used in Ayurvedic medicine for stress. Repels insects, soothes bee stings. *Avoid if pregnant or breastfeeding. Slows blood clotting.*
Basil, Perennial *Ocimum americanum* **OW**	**M–L** Mint. White or purple flowers. Aromatic.	**Short-lived Frost-tender Perennial**	**6–24 in (15–60 cm)** See sweet basil. Suitable for containers and herbal borders.	**BB, HB, SB** The best basil for attracting bees, especially planted en masse.	Edible flowers and leaves brewed into a tea used for treating colds and other ailments. *Avoid if pregnant or breastfeeding. Slows blood clotting.*
Basil, Sweet *Ocimum basilicum* **OW**	**M** Mint. White, pink or light-purple flowers. Aromatic foliage is green or purple.	**Tender Annual**	**1–2 ft (30–60 cm)** Full sun in well-drained moist soil. Start indoors; transplant late spring. Loves hot conditions. Pinch back to bush out.	**BB, HB, SB** Honey potential: 45–90 lbs/ac. Blooms support bumblebee and honeybee health.	Edible flowers and leaves. Specific cultivars for Italian or Thai cuisine. Make pesto. Use in cocktails. *Slows blood clotting and lowers blood pressure.*
Bee Balm, Lemon *Monarda citriodora* **NW**	**M–L** Mint. Spikes with whorls of pink-mauve flowers in tiers. Showy purple bracts. Lemon scent.	**Perennial** Hardy to zone 4. Deer resistant.	**2–4 ft (0.6–1.2 m)** Prefers sun and grows in rich or sandy soil; prefers high clay content. Dry to moderate conditions. Sow seed in early spring.	**BB, HB, SB, BI** Mining, plasterer bees and more. Tiers of flowers accessible to many kinds of bees.	Lemon-scented foliage and flowers used in sweet and savoury dishes. Tea soothes respiratory ailments. Mosquito deterrent. *Avoid if pregnant or breastfeeding.*

*Each pound per acre is equal to approximately 1.1 kilograms per hectare.

PLANT NAME	PLANT FAMILY, BLOOM, FOLIAGE	HARDINESS	HEIGHT AND GARDEN NOTES	BENEFITS TO BEES*	PERKS FOR PEOPLE
Bee Balm, Scarlet, a.k.a. Scarlet Bergamot *Monarda didyma* **NW**	**M–L** Mint. Red or white heads of long double-lipped corollas. Aromatic.	**Perennial** Hardy to zone 4. Deer resistant.	**2–5 ft (0.6–1.5 m)** Partial sun, loamy soil. Moist conditions. Sow in early spring; light required for germination.	**BB, SB** Butterflies, long-tongued bumblebee guilds and queens, hummingbirds.	Edible flowers and leaves used in fruit salads and tea. Makes sweet syrup. Infuse in honey after drying. *Avoid if pregnant or breastfeeding.*
Bee Balm, Wild *Monarda fistulosa* **NW**	**M–L** Mint. Single cluster of pink to mauve flowers, long tubular corolla, protruding stamens. Citrus scent.	**Perennial** Hardy to zone 4. Deer resistant.	**2–4 ft (0.6–1.2 m)** Sun to part shade in sandy to loamy soil. Dry to moist conditions. Start seed indoors; plant out when soil is warm. Spreads by rhizomes. Divide every 2–3 years. Blooms second year.	**BB, HB, SB** Cuckoo, green sweat, leafcutter, long-horned, small resin, sweat, wool carder bees. Nectar robbing. Specialist black sweat bee, *Monarda dufourea*.	Edible leaves and flowers used in tea, added to fruit salads. Essential bee garden herb. *Avoid if pregnant or breastfeeding.*
Borage *Borago officinalis* **OW**	**E–M** Borage. Blue star-shaped bloom, black stamens. White flowers: *B. officinalis* 'Alba'.	**Self-seeding Annual**	**2 ft (60 cm)** Prefers sun. Poor soil is fine if good drainage. Direct-sow. Long taproot. Water well to keep up nectar supply.	**BB, HB** Top honey plant: 200 lbs/ac honey and 60–160 lbs/ac blue-grey pollen. Nectar all day long.	Leaves and flowers make a calming tea. Blossoms pretty as edible flowers or frozen in ice cubes for cocktails. *Avoid if pregnant or breastfeeding, or if you have liver disease. Seeds contain toxins.*
Calendula, a.k.a. Pot Marigold *Calendula officinalis* **OW**	**E–M–L** Aster. Each flower head composed of tiny florets. Centre/ray florets orange. Fragrant foliage. Avoid double varieties.	**Long-blooming Annual**	**1–2.5 ft (30–76 cm)** Sun or partial shade in average, well-drained soil. Deep taproot. Direct-sow. Provide good air circulation around plants to prevent powdery mildew.	**BB, HB, SB, BI** Cuckoo, green metallic, leafcutter, mason, small carpenter, sweat, wool carder bees. Open-access for short-tongued bees.	Edible petals used in salads and teas. Flowers used to make gentle skin-care products. *Avoid if pregnant or breastfeeding.*
Catmint, Japanese *Schizonepeta tenuifolia* **OW**	**M–L** Mint. Long spears of bilabial mauve flowers. Very aromatic.	**Annual** Deer resistant.	**1–2 ft (30–60 cm)** Full sun in average well-drained soil. Direct-sow in spring when the soil is warm, barely covering seed.	**BB, HB, SB, BI** Nectar rich. Medicinal bumblebee plant.	Infusions used topically to treat skin ailments, itchy skin. Decoctions used to treat colds. *Avoid if pregnant or breastfeeding, or if you have liver disease.*
Catnip *Nepeta cataria* **OW**	**M–L** Mint. Whitish flowers, purple nectar guides, long stamens. Aromatic.	**Perennial** Hardy to zone 3. Deer and drought resistant.	**3 ft (90 cm)** Full sun in average well-drained limey soil with high pH. Sow seed spring or fall, cover very lightly.	**BB, HB, SB, BI** Honey potential: 130 lbs/ac. Cuckoo, digger, leafcutter, sweat bees.	Lemon catnip (*N. cataria* ssp. *citriodora*) makes lemony calming tea. Repels insects. *Avoid if pregnant or breastfeeding. Avoid combining with anaesthetic.*
Chives *Allium schoenoprasum* **NW/OW**	**E–M** Amaryllis. Small umbels of pink flowers in spring and summer second bloom. Oniony scent.	**Perennial** Hardy to zone 3. Deer resistant.	**6–12 in (15–30 cm)** Sun to light shade in well-drained soil. Sow fresh seed when soil can be worked. Cut to encourage fresh growth and blooms.	**BB, SB, HB, BI** Honey potential: 40–80 lbs/ac. Ample nectar if watered well, pollen.	Edible flowers and leaves used as garnish for savoury dishes such as eggs, potatoes. Flowers delicious pickled.

PLANT NAME	PLANT FAMILY, BLOOM, FOLIAGE	HARDINESS	HEIGHT AND GARDEN NOTES	BENEFITS TO BEES*	PERKS FOR PEOPLE
Chives, Garlic *Allium tuberosum* **OW**	**M** Amaryllis. Small, white star-shaped flowers in loose umbels. Oniony foliage and flowers.	**Perennial** Hardy to zone 3. Deer resistant.	**1–1.5 ft (30–46 cm)** Sun to light shade in rich, well-drained soil. Sow seed in spring when soil warms. Harvest by cutting at base once leaves are 6 in (15 cm).	**BB, HB, SB, BI** Shallow flowers provide easy-access nectar and pollen.	See chives. Also, freeze chopped leaves in water in ice-cube trays. Traditional in miso soup, Chinese dumplings.
Cilantro/ Coriander *Coriandrum sativum* **OW**	**M–L–SUC** Carrot. Lacey umbels of small white flowers.	**Self-seeding Annual**	**1.5–2 ft (46–60 cm)** Sun to light shade in well-drained soil. Average moisture. Seed every 2 weeks for successional bloom and harvest. Allow some plants to self-seed.	**HB, SB, BI** Honey potential: 201–500 lbs/ac. Pollen: 100–150 lbs/ac. Plasterer, mining, sweat bees, swallowtail butterfly.	Mature seeds used as spice. Leaves, flowers, roots delicious in Mexican and South Asian cuisines. Use flowers as a garnish for squash soups. *Avoid if pregnant or breastfeeding. May lower blood pressure and blood sugar.*
Dragonhead *Dracocephalum moldavica* **OW**	**M** Mint. Spikes of purple flowers. Lemony foliage.	**Annual**	**1.5–2 ft (46–60 cm)** Sun to part shade in well-drained soil. Start indoors or sow directly in the garden once soil has warmed up, barely covering seed.	**BB, HB, SB** Honey potential: 180–445 lbs/ac. Sugar: 20–30 percent. Ample nectar if watered well.	Edible blossoms. Leaves used fresh or dried. Substitute for lemon balm in recipes. Makes spirit-lifting tea.
Fennel *Foeniculum vulgare* **OW**	**M** Carrot. Dramatic umbels of bright-yellow flowers on tall stalks. Licorice-scented ferny foliage.	**Perennial** Hardy to zone 5. Deer and drought resistant.	**Up to 8 ft (2.4 m)** Sun in well-drained soil. Sow seed in spring after frost. Long taproot. Weedy if self-seeds. Stubble cut in fall.	**BB, HB, SB, BI** Honey plant. Mining, masked bees. Dried stems make good homes for tunnel-nesting bees.	Edible seeds and foliage with a licorice flavour tasty with fish, as tea. Add seeds to baking. Pollen used in Italian cuisine. *Avoid if pregnant or breastfeeding, or if you have a hormonal condition.*
Hyssop *Hyssopus officinalis* **OW**	**M–L** Mint. Purple blossoms. Extend bloom with 'Blue Nectar' and 'White Nectar' varieties. Minty camphor scent.	**Perennial** Hardy to zone 3.	**1–2 ft (30–60 cm)** Full sun, well-drained fertile soil. Start seeds inside or in cold frame. Dig up root ball in spring every few years to remove woody crown. May need staking.	**BB, HB, SB, BI** Honey potential: 80–400 lbs/ac. Funnel-shaped flowers allow access to shorter-tongued bees.	Tea made from leaves/ flowers used to treat colds or indigestion. Use in fruit salads. Stew flowers to make sweet syrup. *Avoid if pregnant or breastfeeding, or if you have epilepsy.*
Lavender *Lavandula* spp. or *Stoechas* spp. **OW**	**M–L** Mint. *Lavandula:* clusters of purple or white flowers with short tubes, purple calyxes. *Stoechas:* purple or white flowers, showy rays. Highly fragrant.	**Long-blooming Perennial** Hardy to zone 5–8 depending on species. Deer and drought resistant.	**1–4 ft (0.3–1.2 m)** Prefers hot, sunny sites with excellent drainage. Poor soil is fine. Top dress annually with compost. Shear back plants by one-third in early fall, avoiding woody stems. Easy to propagate from cuttings.	**BB, HB, SB** Honey potential: 70–120 lbs/ac. Pollen: 250–300 lbs/ac. Leafcutter, mason, wool carder bees. Ample nectar if watered well in dry periods.	Put a few buds in sugar for baking cookies and cakes. Lavender bathtub tea is soothing and calming. Buds can be added to tea blends to add flavour and promote relaxation. *Prepubescent boys should avoid. Avoid combining with anaesthetics.*

PLANT NAME	PLANT FAMILY, BLOOM, FOLIAGE	HARDINESS	HEIGHT AND GARDEN NOTES	BENEFITS TO BEES*	PERKS FOR PEOPLE
Lemon Balm *Melissa officinalis* **OW**	**M–L** Mint. Tiny white flowers, blooming a few at a time. Lemon scent.	**Perennial** Hardy to zone 5. Deer and drought resistant.	**2–3 ft (60–90 cm)** Sun to partial shade Start indoors or direct-sow. Seeds need light to germinate. Cut back to prevent self-seeding. *Can be an invasive spreader.*	**BB, HB, SB** Honeybees cannot access the nectar unless the flower is full, so water well to increase flow.	Aromatic leaves used fresh in calming tea, soups, salads, stews. Dry for winter tea.
Lovage *Levisticum officinale* **OW**	**M** Carrot. Umbels of yellow-green flowers on tall stalks. Strong celery flavour and scent.	**Perennial** Hardy to zone 3. Deer and drought resistant.	**Up to 8 ft (2.4 m)** Sun in rich, well-drained soil. Propagate by seed or root division. Long taproot. Start seed inside in early spring or direct-sow in fall.	**BB, HB, SB** Honey plant. Mining and other short- or medium-tongued bees. Swallowtail butterfly host.	Aromatic leaves used sparingly add depth to soups, stews, egg dishes. Large similar to celery seeds to baking. *Avoid if pregnant or breastfeeding. May raise blood pressure.*
Mignonette *Reseda odorata* **OW**	**E–M–L–SUC** Reseda. Fragrant white, red, orange or yellow diminutive flowers on short to long spikes. Sweet scent.	**Annual**	**6 in–1 ft (15–30 cm)** Sun to dappled shade, moist sandy to clay loam. Start outdoors, barely covering seeds. Keep moist until germination. Pinch back. Fertilize to boost scent.	**BB, HB, SB** Leafcutter, polyester bees. Beloved by honeybees. Pumps out nectar all day long when weather is warm.	Young leaves eaten raw in salads. Extremely fragrant flowers dried for potpourri.
Mint *Mentha* spp. **Peppermint** *Mentha × piperita* **Spearmint** *Mentha spicata* **OW**	**M** Mint. Spikes of small, white bilabial flowers. Minty fragrance. Can be invasive, but contained in pots.	**Perennial** Hardy to zone 4. Deer resistant.	**6 in–3 ft (15–90 cm)** Sun to dappled shade in fertile soil. Start seeds indoors 2 months before first frost. Needs snow cover or mulching in harsh winters. Renew by dividing and thinning.	**BB, SB, HB, BI** Honey potential: 150–200 lbs/ac. Leafcutter bees, resin bees, sweat bees.	Fresh or dried leaves make refreshing tea. Can be frozen in ice-cube trays. Freshens breath and aids digestion. Use in salads, baking, desserts.
Motherwort *Leonurus cardiaca* **OW**	**M–L** Mint. Small, pink bilabial flowers with purple spots.	**Perennial** Hardy to zone 4.	**3 ft (90 cm)** Sun to partial shade. Thrives in moist soil. Seeds benefit from cold, moist stratification. Sow early spring or fall.	**BB, HB, SB** Honey potential: 180–445 lbs/ac. A top herb for honeybees.	Medicinal herb used as a sedative and support for women's health. *Avoid if pregnant or breast-feeding, or if you have a heart condition. Can spread vigorously.*
Nasturtium *Tropaeolum majus* **NW**	**M–L** Nasturtium. Flower colours vary from cream to yellow to orange or red. Choose varieties with nectar spurs.	**Annual**	**1–2 ft (30–60 cm)** Sun to partial shade; average, slightly acidic soil. Water if dry to keep roots cool. Sow indoors late winter or direct-sow after last frost.	**BB, HB, SB** Long-tongued bees; smaller bees lap up nectar at edges of spurs. Loved by hummingbirds.	Edible flowers lend spicy flavour to dishes and can be blended into cream cheese. Green seeds pickled like capers. *Avoid if you have ulcers or kidney disease.*

Stay calm with bee balm: bergamot draws parasitic wasps as well as short-tongued bees, making it a good companion for any plant that attracts aphids.

PLANT NAME	PLANT FAMILY, BLOOM, FOLIAGE	HARDINESS	HEIGHT AND GARDEN NOTES	BENEFITS TO BEES*	PERKS FOR PEOPLE
Nigella, a.k.a. Black Cumin *Nigella sativa* **OW**	**M–SUC** Buttercup. Blue or white flowers; long stamens. Feathery foliage allows light on neighbouring plants. Aromatic.	**Self-seeding Annual**	**8–12 in (20–30 cm)** Sun in light, well-drained soil. Direct-sow at 2-week intervals in spring for extended bloom. Light can inhibit germination.	**BB, HB, SB** Nectar produced by glands of modified petals, accessible to bees able to lift the "lid" on nectar pits.	Pungent seeds used in Middle Eastern and Asian cuisine in cheese, curries, naan. Petals edible. Seed used for potent herbal medicine. *Use in small amounts. May lower blood pressure or blood sugar, slow blood clotting, make drowsy.*
Oregano, Common *Origanum vulgare* **OW**	**M–L** Mint. Clusters of small flowers on woody stalks. Blossoms can be pink, white or purple. Aromatic.	**Long-blooming Perennial** Hardy to zone 4. Deer resistant.	**8–32 in (20–81 cm)** Sun in well-drained spot, sheltered in colder zones. Easy to germinate. Pinch back. Grows well in containers, boulevards, borders.	**BB, HB, SB, BI** Honey potential: 90–180 lbs/ac. Sugar: 76 percent. Leafcutter, long-horned, mason, small carpenter, small resin, sweat, wool carder bees.	Leaves of *O. vulgare* lack flavour but fantastic bee plant. Dried and infused with honey, makes cough syrup. Oregano honey is dark, mineral-rich. *May lower blood sugar and increase risk of bleeding.*
Oregano, Greek *Origanum vulgare hirtum* **OW**	**M–L** Mint. White flowers. See common oregano.	**Perennial** Hardy to zone 5.	**8–16 in (20–41 cm)** See common oregano. Sow indoors late winter or direct-sow after last frost.	**BB, HB, SB, Bi** See common oregano.	The best oregano for meat, bean and vegetable dishes. *May lower blood sugar and increase risk of bleeding.*
Poppy, Oriental *Papaver orientale* **OW**	**E** Poppy. Variety of colours and patterns.	**Perennial** Hardy to zone 3.	**2–3 ft (60–90 cm)** Sun, in well drained, moderately fertile soil. Seeds need light to germinate. Direct-sow Feb–March in mild zones, April–May in colder. Long taproot.	**HB** Blue-black pollen. Pollen: 80–120 lbs/ac. Good for filling the "June gap" for honeybees.	Dried pods are decorative. Add seeds to baking.
Rosemary *Rosmarinus officinalis* **OW**	**E** Mint. Blue or white flowers. Highly aromatic.	**Tender Perennial** Hardy to zone 7–8. Deer and drought resistant.	**3–5 ft (0.9–1.5 m)** Sun, in sandy, well-drained soil. Good air circulation helps prevent powdery mildew. Propagate by divisions or cuttings. Re-pot in spring.	**BB, HB, SB, BI** European honey plant. Long-horned, mason, mining, small carpenter, sweat, wool carder bees.	Essential savoury herb in Greek cuisine. Toss chopped leaves on vegetables or meat before cooking. Candy the flowers. *Avoid if pregnant or you have a digestive disorder.*
Sage, Clary *Salvia sclarea* **OW**	**E** Mint. White to light-purple bilabial blossoms open as if yawning. Aromatic.	**Biennial/ Short-lived Perennial** Hardy to zone 4. Deer resistant.	**3–4 ft (0.9–1.2 m)** Full sun in well-drained soil. Sow seeds as wildflowers in early spring or late fall.	**BB, HB, SB** Honey potential: 90–180 lbs/ac. Stamens deposit pollen on the backs of long-tongued bees.	Helps fix the scent of aromatic herbs in potpourri.
Sage, Culinary *Salvia officinalis* **OW**	**M** Mint. Purple bilabial flowers with a deep corolla. Aromatic.	**Perennial** Hardy to zone 5. Deer and drought resistant.	**1.5 ft (46 cm)** Prefers a sunny, well drained site. Sow indoors late winter or direct-sow after last frost.	**BB, HB, SB** Honey potential: 45–445 lbs/ac. Primarily visited by bumblebees, wool carder bees.	Use in poultry stuffing. Fried leaves delicious. Infuse dried leaves and flowers in honey to make cough syrup. *Avoid during pregnancy. May slow blood clotting.*

PLANT NAME	PLANT FAMILY, BLOOM, FOLIAGE	HARDINESS	HEIGHT AND GARDEN NOTES	BENEFITS TO BEES*	PERKS FOR PEOPLE
Savory, Summer *Satureja hortensis* **OW**	**M** Mint. Small pink, white or purple bilabial flowers. Aromatic.	**Annual** Deer and drought resistant.	**6–20 in (15–50 cm)** Sun in rich, well-drained soil. Sow in spring. Sow indoors late winter or direct-sow after last frost.	**BB, HB, SB** Traditional honeybee herb. Honey potential: 45–90 lbs/ac.	Crush and rub on bee stings to reduce pain and swelling. More delicate flavour than winter savory. *Avoid during pregnancy. Avoid ingesting with any medication. May slow blood clotting.*
Savory, Winter *Satureja montana* **OW**	**M** Mint. Small white or purple bilabial flowers. Aromatic.	**Perennial** Hardy to zone 5. Deer and drought resistant.	**6–20 in (15–50 cm)** Sun in well drained sandy soil. Sow indoors late winter or direct-sow after last frost.	**BB, HB, SB, BI** Traditional honeybee herb, even superior to summer savory. Honey potential: 45–90 lbs/ac.	Tea from leaves may aid digestion. Add to beans, poultry, pork. Contains geraniol and used to treat bee stings. *Avoid during pregnancy. Avoid ingesting with any medication.*
Thyme, French or English *Thymus vulgaris* **OW**	**M** Mint. Grey-green foliage. See wild thyme. Invasive in parts of US.	**Perennial** Hardy to zone 4. Deer and drought resistant.	**1–1.5 ft (30–46 cm)** See wild thyme.	**BB, HB, SB** Honey potential: 445 lbs/ac. See wild thyme.	French varieties sweeter than English. See wild thyme. *Avoid during pregnancy, breastfeeding, female hormonal conditions. May interfere with blood clotting.*
Thyme, Wild *Thymus praecox* **OW**	**M** Mint. Small purple flowers. Dense foliage releases a spicy fragrance similar to oregano. Aromatic. Invasive in parts of US.	**Perennial** Hardy to zone 4. Deer and drought resistant.	**6–12 in (15–30 cm)** Sun to part shade, well-drained soil. Mix tiny seeds with sand; sow on prepared soil surface in spring. Divide in early spring and fall.	**BB, HB, SB, BI** Honey potential: 50–150 lbs/ac. Pollen: 200–250 lbs/ac. Cuckoo, leafcutter, mason, mining, resin, small carpenter, sweat, wool carder bees.	Useful and healthful herb for cooking with beans, vegetables, poultry, fish and meat. *Avoid during pregnancy, breastfeeding, female hormonal conditions. May interfere with blood clotting.*
Toothpick Weed, a.k.a. Honeyplant *Ammi visnaga* **OW**	**M** Carrot. White umbels with tiny white flowers attract clouds of pollinators.	**Biennial or Annual in Cold Zones**	**3 ft (90 cm)** Sun in moist but well-drained soil. Seeds benefit from cold, moist stratification, or sow in fall. Long taproot.	**BB, HB, SB, BI** Honey plant. Nectar-rich buffet for many bees and beneficial insects.	Seeds used in Ayurvedic medicine to treat muscle spasms. *Avoid if pregnant or on heart medicine. Sap irritates skin.*

*Each pound per acre is equal to approximately 1.1 kilograms per hectare.

A tiny masked bee, a honeybee and a yellow-faced bumblebee forage in the flowers of garlic chives. The leaves can be used to make delicious green onion pancakes.

Bee-licious Edible Gardens

CHAPTER FIVE

*T*he desire to grow a backyard vegetable garden to provide fresh seasonal food for the table has swept across North America. These new food plots may have smaller footprints than those of the Victory Garden movement that inspired them, but planted right, they can take big steps toward supporting foraging bees. Yesterday's Victory Gardeners were encouraged to rotate crops to keep the soil healthy and prevent pests and disease, and that remains good advice to this day. Cover crops planted between harvests (such as buckwheat, crimson clover and lacey phacelia) also amend the soil if left to bloom. Meanwhile, growing an extra row of parsley or onions and letting it bolt will have bees ecstatically humming around floral umbels and orbs. And the increasingly popular emphasis on four-season gardening can be a boon to bees in the early months of spring, when overwintering brassicas, carrots and onions bloom.

A small sweat bee (*Halictus*) fills her "jodhpurs" with pollen from kale flowers. Children and bees love to snack on flowers from radish, kale, broccoli and Brussels sprouts.

Growing a succession of crops for harvesting food throughout the seasons parallels the need for plants that provide pollen and nectar for bees through fall and in spring when native bees emerge.

VICTORY TRIOS—PLANT IN THREES FOR THE BEES

Planting a healthy number of plants will attract many kinds of bees, all of which will find a foraging niche and leave every 'Big Beef' tomato and scallopini squash pollinated. Planting in trios has many bee benefits: it provides a variety of shapes of blossoms accessible by many species of bees, supplying them with a succession of pollen and nectar sources without any seasonal gaps. It allows you to make container gardens into works of art with complementary colours and contrasting textures. A successful threesome creates a synergy of plants that attracts beneficial insects, repels pests, suppresses weeds and optimizes soil health.

THREE SISTERS: SCARLET RUNNER BEANS, SQUASH AND SUNFLOWERS

The interplanting of squash, beans and corn has a long tradition among Indigenous societies in the Americas, such as the Anasazi, Ewa, Iroquois and Maya. Sometimes sunflowers are added as a fourth sister, which is a better choice than wind-pollinated corn when it comes to gardening for bees. Squash smothers the ground and repels weeds, while offering up organic matter that is tilled back into the soil, improving its drainage and ability to hold moisture. Beans fix nitrogen in the ground for the heavy feeders.

PLANTING PLAN

The Victory Bee Garden: A Bounty for Backyard Bees

The work to save our bees starts right in your own backyard. Planting nectar-rich fruit-bearing trees and shrubs provides the framework for a simple set of vegetable plots that can be interplanted with flowers for bees and beneficial insects. At the heart of the backyard Victory Garden for Bees is the herb garden, drawing bees in with intoxicating and healing scents. Climbing vines like scarlet runner beans create an efficient use of space in order to pack as much floral density as possible into a biodiverse habitat that grows a rainbow of nutritious foods for the table. Heritage vegetables and perennials raised organically are a part of our Canadian heritage and will give our bees a fighting chance. See a sample Victory Bee Garden plan on opposite page.

Victory for Bees Starts at Home

In 1943, there were 209,200 Victory Gardens in Canada, producing an average 550 pounds of vegetables each.[23] In the US, 20 million gardens grew 8 million tons of food.[24] In the same way that these historical efforts filled the food larders of North American families, modern-day Victory Gardens for Bees can provide "pollen pantries" for bees.

ESPALIERED TAYBERRIES

SKATOON BERRY

ROSEMARY

OREGANO

LAVENDER (ANGUSTIFOLIA)   LAVENDER (INTERMEDIA)   LAVENDER (STOECHAS)   OREGON TEA

SAGE

WINTER SAVORY

SALMONBERRY

THYME

SEDUM

CHERRY TREE

ALPINE STRAWBERRY

SEDUM

STUMP

VIOLET

DINING AREA

THYME

SEDUM

BEE HOTEL (GREEN ROOF)

TIERED HERB GARDEN

SQUILL

SEDUM

COMFREY

BARE SOIL

BORAGE

STUMP

MORNING GLORY

CLEMATIS

BEE OASIS

BLACK CURRANTS

EDIBLE FLOWERS

SALAD GREENS

ROSE

NODDING ONION

COMPOST

AMARANTH

HONEYSUCKLE

BEE MEADOW (CAN BE ROTATED TO A VEGETABLE BED)

ORACH

CUT FLOWERS

RASPBERRY

TOMATOES

PEPPERS

SNOWBERRY

LUPIN

RUNNER BEANS

THYME

ARTICHOKES   MARIGOLD

SUNFLOWER

BROCCOLI

CANADA GOLDENROD

TALL CONEFLOWER

LEEKS

DILL

CARROTS

BUSH BEANS

BRUSSELS SPROUTS

RADISHES

SQUASH

GARLIC

HELENIUM

POTATOES

KALE

BUSH BEANS

POACHED EGG FLOWER

PURPLE CONEFLOWER

ONIONS

BLUEBERRY

MARIGOLD

CALENDULA

CALENDULA

ASTER

PERENNIA SAGE

TABLE   BENCH   TABLE

SOLOMON'S SEAL   COLUMBINE

MASTERWORT

HARDY GERANIUM

MASTERWORT

HARDY GERANIUM

Beans and squash also complement each other nutritionally. Dried beans and sunflower seeds are rich in protein, and squash yields vitamins from the fruit and healthful oil from the seeds. This synergistic trio also supports bees, especially if you substitute sunflowers for the corn and choose scarlet runner beans, which pump out nectar for bumblebees and hummingbirds. Squash bees depend on plants like pumpkins to provide food and lodging. Each large sunflower head has 300 to 400 small florets that provide nectar and pollen for bees. Varieties like 'Earthwalker' are valuable because they have multiple heads that bloom at different times to extend the bloom season. Choose heritage varieties that haven't had their pollen bred out of them.

A VICTORIOUS COMBINATION: DR. CARROT, CHIVES AND RADISHES

Carrots, chives and radishes are allies working together to feed busy families and bees. Carrots were a Victory Garden staple, with popular propaganda posters declaring "carrots keep you healthy and allow you to see in the blackout." "Dr. Carrot" advised children to eat carrots regularly, for breakfast as "carrolade" (carrot lemonade) and even as dessert, served on a stick lollypop-style![25] Save garden space by alternating carrots and radishes in a row: the radishes mature first, loosening the soil for their companions. But do leave a few radishes and carrots in the ground to overwinter and flower for bees. Adding chives to this combination will improve the flavour of the root crops, discourage carrot rust fly and fill in bloom gaps. Experiment with both spring and summer radishes, allowing some to bolt, to further spread out a flow of flowers for foraging bees of all stripes.

SALAD SOULMATES: TOMATOES, BASIL AND LEMON BEE BALM

Basil and bee balm will attract the bumblebees needed to buzz-pollinate your tomatoes to make them voluptuous and tasty. Any of the lemon bee balm and basil species will improve the health and flavour of tomatoes, while offering up nectar for the bumblebees that pollinate the fruits. Bumping up the aromatic herbs by adding lime (perennial) basil will attract more bees and inspire you to create artful Caprese salads with unique flavour combinations.

Opposite, clockwise from top left: Male sunflower specialist bees (*Melissodes agilis*) nestle down for a nap alongside a stranded female honeybee taking advantage of their shared warmth (HILLARY SARDIÑAS PHOTO). Fall at UBC Farm is an exciting time, with an abundance of organic pumpkins and other winter squash in all shapes and sizes. At the Salt Spring Centre of Yoga, nasturtiums and morning glories grow among the greenhouse tomatoes, encouraging buzz-pollinating bumblebees to work the tomato plants. Small sweat bees (*Halictus*) and other species find radish blossoms ravishing.

Squash blossoms provide habitat for bees and the occasional frog at University of British Columbia Farm.

IMPROVED FLOWER POWER: BROCCOLI, LEEKS AND CALENDULA

Purple sprouting broccoli, a vigorous overwintering brassica, is particularly good value for space and money because it grows to a towering height and offers multiple stems with buds and flowers that help bees build strong brood in spring. Leeks produce hundreds of blossoms in a small space and improve the flavour of the broccoli. Calendula repels harmful nematodes and provides a long season of nectar and pollen for short-tongued bees and beneficial insects.

THE BUMBLEBEE BUNCH: MARIGOLDS, BUCKWHEAT AND TOMATILLOS

Marigolds are superheroes in the garden, chasing away whiteflies and killing harmful nematodes. They will help your bee plants thrive, and if you avoid the

MARIGOLD BEE STING

To make infused syrup, place **2 Tbsp (30 mL) dried marigold petals** and **2 Tbsp (30 mL) dried chili peppers** in a clean 1-cup (250-mL) mason jar. Pour **½ cup (120 mL)** honey over top and stir to combine. Seal tightly and allow to infuse on a counter top for 3 weeks, turning twice a day.

For each cocktail, combine **1 Tbsp (15 mL) infused honey** with **1 Tbsp (15 mL) water**. Shake with **2 oz (60 mL) vodka** and **juice from 1 medium lime** in a cocktail shaker (or mason jar) and pour over ice. Garnish with a Mexican sour gherkin.

⟨ **recipes** ⟩

double-blossomed varieties bees will use them in periods of dearth. Marigolds are best buddies with many bee plants, but not with beans or brassicas. Look for resin-scented French marigolds (*Tagetes patula*) and Mexican marigolds (*T. minuta*), which are even stinkier, giving them stronger pest-repelling superpowers. Buckwheat also has superpowers, with summer blooms making up for what marigolds lack in nectar, attracting many species of beneficial insects and the bumblebees that tomatillos need for buzz pollination.

BORDER BUDDIES: PARSNIP, ANISE HYSSOP AND SCARLET BEE BALM

Believe it or not, one of the biggest bee magnets in my garden is an overwintered flowering parsnip! There are salient facts about parsnips you should know. First of all, you need to wear gloves when cutting the stalks of the plants because it oozes a sap caustic to the skin. Do not eat parsnips after they've bloomed, as the roots become a concentrated course of a hallucinogenic chemical called "myristicin," the same one that makes nutmeg mind-altering if you eat too much. Also, this large, long-root crop needs 2 feet (60 cm) of soil worked with a shovel. The best time to harvest parsnips is after first frost. To germinate, seeds need to be very fresh, which is why I like parsnip to self-seed; I've also had success germinating seeds in a plastic bag with a moist piece of paper towel. The nectar-rich blossoms of anise hyssop and scarlet bee balm work together with the umbels of parsnips to create a long-blooming summer picnic for bees.

PICKLING PALS: CUTE CUKES LOVE DILL AND NASTURTIUMS

Mexican sour gherkins (*Melothria scabra*) are so adorable, they are perfect for school gardens. These grape-sized cucumbers have the uncanny appearance of watermelons that have been zapped by a magic shrink-ray. The small yellow flowers are pollinated by small and medium bees. The delicate vines are easy to train up a trellis intertwined with other climbers like sweet peas. It's easy to remember that cucumbers love dill if you love dill pickles. Take a page from a classic Victory Garden and plant a pickling garden with dill and cukes, tossing in nasturtiums with seeds you can pickle like capers to add to your mason jars. This synergistic trio provides nectar for a range of bees, from

I made an apple and chayote pie, and tested it on my son and husband. It received a thumbs-up from my toughest critics.

those with tiny tongues to long-tongued bumblebees and hummingbirds.

GO VERTICAL FOR VICTORY: CHAYOTE, ZUCCHINI AND OREGON TEA

Look up—waaay up—and see how high you can grow your way to victory. In order to pack in more blossom density for bees and extra food for you, create a sturdy trellis for chayote to climb, while the accompanying zucchini plants produce fruit and flowers week after week through the summer heat and right into autumn. What is known as "chayote" in Spanish is called "Buddha's hand melon" in Chinese, "mirliton" in French creole. Grow the chayote from a locally produced vine to ensure it is suitable for your region. The seed can only germinate inside the fruit, which needs to be kept in a dark cupboard over the winter. Plant the fruit in the ground with just its tip above the soil. The small yellow-green flowers are pollinated by bees, wasps and hummingbirds. Choose a compact, high-yielding variety of zucchini for a long succession of blooms for bees. If you can't find chayote, plant your favourite climbing cucumber or squash. Draped over the edge of hanging baskets, Oregon tea makes a sweeping final touch and is a long-blooming source of nectar for thirsty bumblebees.

SPRING, SUMMER AND FALL BLOOMS FOR BEES: ASTER, COSMOS AND KALE

The dark curly foliage of 'Redbor' kale and the rumply texture of dinosaur kale have a dramatic presence in the soft autumn sun as cosmos and asters bloom behind them, their glowing yellow centres providing nectar and pollen for late guilds of bees. New queen bumblebees bulk up for winter on the fall flowers, and

Opposite, clockwise from top left: Kale, leek, zucchini and cosmos blossoms visited by bees. **Right:** A mini sunflower greens garden.

The Queen's Neeps

During World War I, Queen Mary had the gardener replace a perennial border at Buckingham Palace with "an abundance of royal turnips." Following her example, Queen Elizabeth II worked in the "Dig for Victory" garden at Windsor Castle as a teenager during World War II. In 2009, an organic vegetable patch was created at Buckingham Palace as part of a grow-your-own movement inspired by war gardens. Among the heritage-variety crops planted are garlic, leeks and onions to deter pests and feed bees.[26] You can grow your own royal turnips and alliums by leaving some to flower for the queen bees.

in the spring, hibernating queens wake up to a feast of kale blossom nectar, to fuel them in the building of nests and laying of eggs. Chocolate cosmos look stunning cozied up to dark-purple kale; add dahlias and *Verbena bonariensis* to top up the bee benefits.

DIG FOR VICTORY TODAY: FILLING FOOD BANKS FOR BEES

There are Victory Gardens created in World War II that still remain active community gardens to this day. And new Victory Gardens are popping up too, like the Hamilton and Oshawa Victory Gardens in Ontario that grow produce for the local food banks. Victory Gardens 2007+ is a project started by lead artist Amy Franceschini in partnership with the San Francisco Department for the Environment. The goal of this project is to reclaim part of the original Victory Garden in Golden Gate Park and create a network of Victory Gardens around the city. Since 2009 the Great Milwaukee Victory Garden Blitz has created 2,500 gardens in backyards, schools, churches, temples and community centres. If each of these old and new Victory Gardens were geared toward feeding bees as well as humans, this network of community gardens could provide pollinator pathways crisscrossing cities and towns across North America, creating food banks for bees.

COMFREY FOR BEE-FRIENDLY COMPOST

Soil contains three key minerals: nitrogen, potassium and phosphorus. Heavy feeders—potatoes, squash and corn, for example—will benefit from extra nitrogen. Flowers and fruits need phosphorus in the soil. Root vegetables need potassium. One very bee-friendly way to add these minerals to the soil is with compost made from the common weed comfrey (*Symphytum officinale*). Comfrey is a classic bumblebee magnet, attracting long-tongued bees with flowers that copiously refill with nectar every 45 minutes. Comfrey's taproots can dig down as deep as 6 feet (1.8 m) into the soil, mining deep-seated minerals. Added to the compost pile, comfrey breaks down quickly to make a rich soil amender.

Think about packing as much colour as possible into your Victory Garden for Bees with flowers and vegetables that come in a rainbow of hues.

VEGETABLES for Bees

A SEASONAL PLANTING CHART

*G*rowing food for yourself means you can also feed the bees if you let some vegetables blossom and go to seed. By saving these seeds you can help the bees while saving money, ensuring that the seeds you use for next year's crops are fresh, organic and dependable varieties. Extra flowers can be pickled, used in salads, or used as a garnish, adding a rainbow of colours to your culinary creations. Biennial plants usually flower in the second year, so if you can selectively leave some veggies in the ground over the winter, you'll provide bloom for bees early in the next year, when they may need it most.

Plants in this chart need at least six hours of sun in well-drained, fertile soil, and the more sunshine, the better. When planting from scratch, check the seed package for how deep to plant the seed, or follow the rule of thumb: plant a seed at a depth roughly double its diameter. Heights noted for plants are maximum; they will vary according to weather and soil conditions, and blooms/bolting have been taken into consideration, as leaving some plants to bloom for the bees is essential to your Victory Garden for Bees. "Days to maturity" refers to the maximum number of days for food harvest, but please do leave some of the plants in longer for maximum blooming; finished blooms can be snipped off to encourage new blossoms until the plant peters out. Root crops are not good for eating once the plant has flowered, so plant enough for your own harvest and allow some extra plants to bloom for the bees. Spacing of plants refers to the mature plant, not the seed—when sowing seed in the ground, plant extra and thin out, leaving the most vigorous seedlings.

PLANT ORIGIN
OW: Old World (indigenous to Europe, Asia or Africa)
NW: New World (indigenous to the Americas and their islands)

BLOOM PERIOD
E: Early
M: Mid-season
L: Late
SUC: May be seeded in succession

HARDINESS
refers to the coldest zone the plant is hardy to.
DEER AND DROUGHT RESISTANCE are most reliable once the plant is established.

BEES ATTRACTED
BB: Bumblebees
HB: Honeybees
SB: Solitary bees
BI: Beneficial insects

PLANT NAME	PLANT FAMILY AND BLOOM	HARDINESS	HEIGHT AND GARDEN NOTES	BENEFITS TO BEES*	EDIBLE PROPERTIES
Artichoke *Cynara cardunculus* var. *scolymus* **Cardoon** *C. cardunculus* **OW**	E–M Aster. Large, plush thistle-like purple blooms.	**Annual/ Biennial or Perennial** Hardy to zone 8. Deer resistant.	**80 in (2 m)** **Spacing: 3 ft (90 cm)** **Days to Maturity: 180** 2-season crop in warm zones; mulch over winter. In colder zones start indoors January.	BB, HB, SB, BI Bumblebees swim in stamens to drink nectar and collect pollen. Cardoon flowers exceptional for bees.	Artichoke buds and stems. Remove spine of larger fruits. Cardoon buds not as tasty but worth growing.
Bean, Scarlet Runner *Phaseolus coccineus* **NW**	M–L Legume. Pea-like white to scarlet blooms.	**Frost-tender Perennial Grown as Annual**	**8 ft (2.4 m)** **Spacing: 8 in (20 cm)** **Days to Maturity: 55** Plant late spring when ground warms. Sturdy trellis/tripod needed.	BB Bumblebees and hummingbirds love the scarlet blooms, which require trip pollination.	Blooms, pods, seeds. Pick beans when young to cook while green or leave to mature for dried beans.
Broccoli *Brassica oleracea* **OW**	E (overwintered) Brassica. Yellow blooms with four petals.	**Biennial or Annual in Cold Zones**	**2 ft (60 cm)** **Spacing: 20 in (50 cm)** **Days to Maturity: 90** Overwinter for spring harvests and allow full bloom for bees.	BB, HB, SB, BI Blossoms on overwintered stalks vital forage for bees.	Blooms, buds, leaves, stems. Sweetens after frost. Eat a few heads at a time, fresh or cooked.
Brussels Sprouts *Brassica oleracea* **OW**	E (overwintered) Brassica. Yellow blooms with four petals.	**Biennial or Annual in Cold Zones**	**4 ft (1.2 m)** **Spacing: 20 in (50 cm)** **Days to Maturity: 100** Benefits from rich soil. Harvest throughout winter and into spring.	BB, HB, SB, BI The blooms on overwintered stalks are important for bees.	Buds, blooms, leaves. Snip off tiny cabbages after first frost for sweeter taste.
Carrot *Daucus carota* **OW**	M–L (overwintered) Carrot. Lacey umbels of tiny white, pink or burgundy flowers.	**Biennial or Annual in Cold Zones**	**3 ft (90 cm)** **Spacing: 3 in (7.5 cm)** **Days to Maturity: 80** Avoid fresh manure. Dig soil well, keep moist, weeded. Use row cover to deter carrot rust fly; uncover if flowering to give bees access.	BB, HB, SB, BI Honey plant. Cuckoo, leafcutter, mining, sweat, masked bees. Small bees love the dainty umbels of white flowers in the carrot family.	Blooms, roots. Enjoy roots raw or braised, steamed or roasted, in juice and soups. Eat blooms in tempura batter fritters. Compost roots of carrots that have bloomed; do not eat.
Chayote *Sechium edule* **NW**	M–L Squash. Shallow yellow-green male and female blooms on the same vine.	**Tender Perennial**	**Climbing vine** **Spacing: 10 ft (3 m)** **Days to Maturity: 150** Plant in your sunniest spot. Extensive root system; well drained soil essential. Requires sturdy trellis to hold fruit above ground.	BB, HB, SB Adored by small- to medium-tongued bees and hummingbirds. Stingless bees (*Trigona*) native to Mexico to Costa Rica are the native pollinators.	Edible roots, stems, leaves, fruit and flowers. Peeled fruits eaten cooked or raw. Cut-up fruit can be combined with apples to make delicious pies. Use in Chinese cooking.
Cucumber *Cucumis sativus* **OW**	M Squash. Yellow flowers with plush stamens in male flowers.	**Annual**	**Grow as vine or sprawling plant** **Spacing: 20 in (50 cm)** **Days to Maturity: 65** Rich soil and sunshine. Trellis climbers.	BB, HB, SB Cuckoo, large carpenter, leafcutter, mining, small carpenter, squash, sweat bees.	Blooms and fruit. Raw cucumbers eaten sliced with oil and vinegar or yogurt and dill. Some grown for pickles.

*Each pound per acre is equal to approximately 1.1 kilograms per hectare.

PLANT NAME	PLANT FAMILY AND BLOOM	HARDINESS	HEIGHT AND GARDEN NOTES	BENEFITS TO BEES*	EDIBLE PROPERTIES
Gherkin, Mexican Sour *Melothria scabra* **NW**	**M** Squash. Small yellow flowers.	**Annual**	**Grow as vine or sprawling plant** **Spacing: 1 ft (30 cm)** **Days to Maturity: 75** Rich soil and sunshine. Grow on a small trellis.	**HB, SB** Short- to medium-tongued bees.	Fruits and flowers. Use fruit as cocktail garnish, add to salads or make into pickles. Fruits look like tiny watermelons.
Kale *Brassica oleracea* **OW**	**E–M–SUC (overwintered)** Brassica. Abundant yellow shallow blooms.	**Biennial or Annual in Cold Zones** (Red Russian is self-seeding)	**40 in (1 m)** **Spacing: 20 in (50 cm)** **Days to Maturity: 60** Grow purple, black, green cultivars for bloom and striking foliage. Prolific all year in temperate zones.	**BB, HB, SB, BI** Brassicas have honey potential of 22–45 lbs/ac. Spring blooms of overwintered kale important food for mason bees. Avoid ornamental kale, which doesn't provide nectar or pollen for bees.	Buds, blooms, leaves. Snip off for smoothies, soups, pesto. Dehydrate leaves for kale chips. Massage leaves with lemon, olive oil, salt, pepper for salads.
Leek *Allium ampeloprasum* **OW**	**M** Amaryllis. White to pink globes contain up to 3,000 tiny flowers.	**Biennial or Annual in Cold Zones** Hardy to zone 4.	**40 in (1 m)** **Spacing: 6 in (15 cm)** **Days to Maturity: 150** Side dress with fish emulsion, compost or organic fertilizer. To overwinter plant 8 weeks before frost.	**BB, HB, SB** Honey potential: 45–90 lbs/ac. Overwinter some to bloom for spring bees. Some types flower in first year if started indoors.	Blooms and stalks. Eat like green onions, or as leek and potato soup. Use fried greens and tiny blooms as garnish. Wash between plant layers before eating.
Onion, Welsh, a.k.a. Japanese Bunching Onion *Allium fistulosum* **OW**	**E** Amaryllis. Globes of white flowers pretty enough to use as ornamentals.	**Perennial** Hardy to zone 3.	**20 in (50 cm)** **Spacing: 20 in (50 cm)** **Days to Maturity: 120** Snip continually, leaving some to bloom in succession. Cut back to renew after flowering.	**BB, HB, SB, BI** Honey potential: 45–90 lbs/ac. Bumblebees are important pollinators.	Blooms and stalks. Cut stems off at the base, and use as you would leeks. Flowers are delicious in salads, on potatoes, on devilled eggs.
Parsnip *Pastinaca sativa* **OW**	**E–M** Carrot. Umbels of tiny yellow-green flowers.	**Biennial or Annual in Cold Zones**	**5 ft (1.5 m)** **Spacing: 2 ft (60 cm)** **Days to Maturity: 120** Follow planting advice for carrot. Large, long root crops like parsnips need 2 ft (60 cm) deep soil that has been worked.	**BB, HB, SB, BI** Small mining bees and other short-tongued bees. Specialist pollinator *Andrena ziziae*. Black swallowtail butterfly host.	Roots. Coat sliced roots in olive oil, toss with minced garlic, and roast. Use care when handling: leaves and stems have toxic sap that can cause rash. Root eaten first year only; do not eat root from plant that has flowered.
Radish *Raphanus sativus* **OW**	**E–M–L–SUC** Brassica. White cross-shaped flowers with 4 petals, faint purple nectar guides. Invasive in California.	**Annual** (often self-seeding)	**1 ft (30 cm)** **Spacing: 1 in (2.5 cm)** **Days to Maturity: 25** Summer-grown Spanish radishes overwinter in warm zones. Add seed to bee-pasture mixes.	**BB, HB, SB, BI** Honey plant. Highly accessible nectar and pollen. Visited by small bees including small sweat bees.	Entire plant. Roots and flowers are eaten raw in salads. Leaves cooked in soup or sautéed. Braise roots in butter with salt and pepper.

A queen bumblebee forages next to a worker bee in a cardoon
flower. The queen is twice the size of the worker bee.

PLANT NAME	PLANT FAMILY AND BLOOM	HARDINESS	HEIGHT AND GARDEN NOTES	BENEFITS TO BEES*	EDIBLE PROPERTIES
Squash, Summer *Cucurbita* spp. including **Zucchini** *Cucurbita pepo* NW	M-L Squash. Large yellow to orange blooms that envelop the bees. Plants feature male and female blooms.	Annual	**Grow as vine or sprawling plant** **Spacing: 2 ft (60 cm)** **Days to Maturity: 60** Plant 1 month after last spring frost. Full sun, rich, evenly moist, well-drained, amended soil. Support climbers with strong trellis.	**BB, HB, SB** Some nectar, copious pollen. Specialist squash bees: *Peponapis* and *Xenoglossa.* Bumblebees and long-horned bees native pollinators of squash.	Blooms and fruit. Fruits can be eaten raw or cooked in many ways, even used in baking. Remove pistils and stamens to use flowers in lasagne or risotto, or fry in tempura batter.
Squash, Winter *Cucurbita* spp. including **Pumpkin** *C. pepo var. pepo* **Butternut Squash** *C. moschata* NW	M Squash. See summer squash.	Annual	**Grow as vine or sprawling plant** **Spacing: 4 ft (1.2 m)** **Days to Maturity: 95** See summer squash.	**BB, HB, SB** See summer squash.	Blooms, fruit and seeds. Eat flesh in soups and stuffed pasta, traditional sweet pumpkin pies. Roast seeds with salt. Cook flowers as with summer squash.
Sunchoke, a.k.a. Jerusalem Artichoke *Helianthus tuberosus* NW	M-L Aster. Classic yellow flowers attract sunflower specialist bees. Tubers can spread. Choose sterile cultivars.	Perennial Hardy to zone 4.	**80 in (2 m)** **Spacing: 20 in (50 cm)** **Days to Maturity: 150** Install a barrier 2 ft (60 cm) deep to contain growth. Full sun, loose, well-drained soil. Plant tubers in spring when soil can be worked.	**BB, HB, SB, BI** Fall honey plant. Cuckoo, digger, leafcutter, mining, sweat bees. Specialist bees *Melissodes agilis* and *Diadasia enavata.*	Roots. Rhizomes eaten cooked like potatoes or raw. Made more digestible by pickling; use pickled slices on risotto or in sushi.
Tomatillo *Physalis ixocarpa* NW	M Nightshade. Yellow bloom, black centre, long stamens.	Tender Annual	**30 in (76 cm)** **Spacing: 15 in (38 cm)** **Days to Maturity: 80** Full sun, rich soil. Plant when soil warms. May need support.	**BB, HB, SB** Plants require buzz pollination.	Fruits. Husked fruits eaten raw or cooked in salsas and other Mexican dishes.
Tomato *Solanum lycopersicum* NW	M Nightshade. Yellow star-shaped blooms, protruding stamens.	Tender Annual	**Determinate 40 in (1 m); Indeterminate: 10 ft (3 m)** **Spacing: 2 ft (60 cm)** **Days to Maturity: 85** Full sun, rich soil. Vining plants need support; pinch back.	**BB** Plants require buzz pollination.	Fruits. Immature green tomatoes can be eaten fried or in mincemeat and salsa. Mature fruits used in salads or cooked. Grow and eat with basil.
Turnip *Brassica rapa* subsp. *rapa* OW	E (overwintered) Brassica. Yellow bloom with 4 petals.	Biennial or Annual in Cold Zones	**30 in (76 cm)** **Spacing: 20 in (50 cm)** **Days to Maturity: 60** Rich soil, loosened to 15 in (38 cm). Do not overwater or roots become bitter: 1 in (2.5 cm) of water per week.	**BB, HB, SB, BI** Blooms on overwintered stalks are important spring food for bees. Compost roots on plants gone to seed; they shouldn't be eaten.	Blooms, leaves and roots. Sauté leaves and blooms with garlic and chili flakes. Roots harvested as baby vegetables in summer or as mature roots. Compost roots of turnips that have bloomed; do not eat them.

*Each pound per acre is equal to approximately 1.1 kilograms per hectare.

Wild for Indigenous Bees

CHAPTER SIX

North American native bees have been undergoing an alarming decline in the last 30 years, with one example being a devastating 50 percent loss of native bee species from their historical habitats in the midwest in the last century.[27] As their populations dwindle, it is increasingly clear how important native bees are to food security and biodiversity. More efficient pollinators than honeybees, they are doing more with fewer resources and absolutely free of charge. With pollination deficits for food crops on the rise, the least we can do is encourage this mighty workforce with room and board and a safe place to raise offspring. Protect habitat, protect the bees. Create habitat, save more bees.

DIY NESTING SITES FOR NATIVE BEES

Go ahead and think of your whole yard as a "Bee & Bee," a potential hotel and nursery for native bees. In addition to following the 16 rules of bee-safe gardening (page 12)—particularly vital for *native* bees—use these 5 tactics to help to create housing for vulnerable pollinators.

1. DESIGNATE A BARE-SOIL BEETOPIA

What you think of as the ugliest and scruffiest part of your lawn may be the best place around in the eyes of ground-nesting bees. Put up a sign to declare it their "Beetopia" and give up worrying about mowing this bit. Try to leave soil in each area of your garden where conditions are slightly different so that each species of bee can choose what suits it best. Include a sloped site with good drainage.

Opposite: My neighbour Costa grows a variety of native coneflowers from seed, layering the tall coneflowers with brown-eyed Susans.

A digger bee (*Anthophora*) emerges from a nesting hole. Bee watchers can learn to identify the signs of ground-nesting bees: little mounds of soil around small, pencil-sized holes.
KATHY KEATLEY GARVEY PHOTO

A joule is a measure of energy, which is what bees collect in nectar, equal to 1 newton of force applied to an object for the distance of 1 metre.

2. SUPPLY NESTING MATERIAL

If bees are going to set up home in your garden, they need nesting materials. Some resources may come with the nesting site, such as the wood that carpenter bees chew to make pulp for building cell divisions. Leafcutter bees require leaves that are smooth on one side and not too thin or too thick, to make good walls for their nests—rose, bean or snowberry leaves will do the trick. Resin bees need pine logs to gather resin for lining their nests. Mason bees hunt for wet mud high in clay; you can put some out in shallow containers or simply make a deep wedge in the soil with a shovel in a shady section of your garden. Polyester bees look for plants with linalool to make the secretions to line their homes—grow cilantro, lavender, monarda and basil to provide this bee-supporting chemical.

3. MAKE A STUMPERY

A stumpery is a collection of rotting wood stumps placed artfully in a disused, shady part of the garden to create fungus-filled habitat and provide housing materials for beneficial beetles, butterflies and, of course, bees. Make a circle of stumps or logs and fill it with leaf mulch to make a cozy hibernaculum for queen bumblebees and habitat for butterfly cocoons.

PLANTING PLAN

A Shade Garden: Wild Plants for Native Bees

While bees generally work better in sunny conditions, some native bees—such as blue orchard mason bees and bumblebees—have the ability to forage in lower temperatures and are more likely than honeybees to seek nectar and pollen in shade. Plants with long nectar spurs, such as lungwort, Dutchman's breeches and wild columbine, are examples of native bumblebee forage found in dappled forest edges and clearings. Shade gardens are also the ideal setting for rotting stumps that provide nesting sites and fungal benefits for bees[28] and serve as hibernacula for bumblebee queens that have gorged on nectar from fall flowers. See a sample shade garden plan on opposite page.

Give tunnel-nesting bees a helping hand by drilling a few holes of varying sizes in the decaying wood. Tunnels of 0.25 inches (6 mm) or less in diameter should be 3 to 5 inches (7.5 to 12.5 cm) deep. Those larger than 0.25 inches (6 mm) in diameter should be 5 to 6 inches (12.5 to 15 cm) deep. If you're pressed for space and don't have room for stumps, use small log rounds, untreated lumber, tree roots and large pieces of bark.

4. CREATE HOLLOW-STEM HOTELS

Bees hollow out small twigs to create tunnels to lay eggs inside, and the easiest way to create a bee hotel is to give plants a stubble cut in the fall. Don't cut stems right down to the ground; instead, leave enough for bees to nest in. Thick stems are best left at least 1 foot (30 cm) long; thinner stems should be about 8 inches (20 cm).

Opposite, clockwise from top left: When creating a pollinator hotel, it is important to provide stems and drill with a variety of diameters. The pollinator hotel in Vancouver created by Nathan Lee and Hartley Rosen is attracting tenants. Bee hotels made from log rounds. On a busy day at the bee condo, three female blue orchard mason bees provision their nests.

To create further nesting opportunities for bees, house the cuttings in a waterproof structure firmly holding them in place. It's very important that the stems don't get knocked about while bees are developing, so keep them protected, off the ground and away from rodents, skunks and racoons, which will naturally take one look at your bee hotel and think it's a vending machine. Clay drainage cylinders work well to house stems, along with large tin cans and wooden cubbyholes. Orient the stems horizontally, with the openings toward the light and the other end against the wall of the hotel. The best stems for bee hotels are either pithy with spongy tissue, such as those found in the rose family, or hollow like Joe-Pye weed, large sunflowers, straw, dry reeds, lovage, fennel, bamboo and some grasses.

5. MAKE ROOM FOR GENTLE BLUE ORCHARD MASON BEES

Sweet and somewhat timid, blue orchard mason bees (*Osmia lignaria lignaria* and *O. lignaria propinqua*)

are wonderful guests in your garden. The female has a mild defensive sting she rarely uses. She drinks nectar for her own energy and creates bee bread for her progeny by mixing regurgitated nectar from her honey stomach with pollen. You may also see her gathering mud in her jaws to make cell walls; it can take her more than 50 trips just to make the little mud wall between brood chambers.

A female mason bee that has mated successfully can lay both male and female eggs, starting with females in the back of the tunnel and then males at the front. Female cocoons are much larger than male.

Female mason bees are highly efficient pollinators, visiting 17 blossoms per minute and needing 850 blossoms to provision for each egg.

If you have fruit or nut trees in your yard that need pollinating in spring, putting up at least one condo for blue orchard mason bees provides big payback, with these eager pollinators greatly improving the quality and quantity of harvests. Your close neighbours will also reap the benefits. You may already have local mason bees happy to move into the home you provide. Look for them in cherry trees, Oregon grape or flowering kale from April to June. If you see mason bees within about 300 feet (90 m) of your home, count yourself lucky! It's very likely they will be happy to set up house in your mason-bee condo. If you don't spot any, you can purchase cocoons from a reputable source.

A mason bee female carries mud with her mandibles to build nest walls. Providing mason bees with fresh mud can save them time and energy. ANNA HOWELL PHOTO

The Cocoon Stage

During the summer, each egg hatches, turns into a larva and finally pupates inside a small grey cocoon. You may have cocoons in the habitats you have provided for the native mason bees in your backyard, or you may have bought them from a local mason-bee keeper, or both. Purchased cocoons, or cocoons that you have retrieved from your yard for cleaning (see Why Clean Mason-bee Cocoons? on page 112), should be stored in the fridge, where they will remain dormant until there are enough spring blooms to feed them. Keep an eye on the weather and look for buds on overwintering brassicas, *Pieris japonica* and apple and cherry trees. Take a walk up and down the block and ask yourself, "Is there a critical amount of blossoms to support these bees?"

It's a very special day when the blooming is sufficient and the weather balmy enough for mason bees to emerge and explore your yard. Take the cocoons out of the fridge and place them outside. Don't worry—you don't need to put them back into the tunnels! Simply place the cocoons near the condo in a weather-proof container with one dime-sized hole on the front. Some condos have cavities inside specifically for releasing cocoons.

Artist Sarah Peeble's Audio Bee Booth allows viewers to hear the bees working in the tunnels and see them at work. ROB CRUIKSHANK PHOTO [1]

Providing a Mason-bee Condo

In wet climates, wood is better than plastic to use as a nesting material for mason bees because the wood can breathe and help prevent cocoon rot. Purchase an easy-to-clean mason-bee condo or make one yourself. Condos come in all shapes and sizes: teardrop, A-frame and simple rectangular boxes closed at the back and open to the light at the front.

If you build your own, the ideal diameter for each

Blanket flower (*Gaillardia*) is long-blooming, drought tolerant and easily accessed by many varieties of bees. So let's plant more of it!

Gentle mason bees are extremely efficient pollinators, out-performing honeybees on spring fruit trees. MARK MACDONALD PHOTO

the condo needs to face southeast in an outdoor location that receives good morning light but is sheltered from rain and harsh sunlight. And placing it on a fence or wall rather than a pole will help mason bees find their way home. Adding contrasting patterns to the front of the condo will also help bees locate their entrance. Use nontoxic paint, and avoid painting the insides of the tunnels. Another option is to toast the front of the condo with a torch to emphasize the wood grain. Nesting bees become confused when major landmarks are moved, so try to not rearrange the furniture or garden ornaments around the condo when mason bees are active.

Why Clean Mason-bee Cocoons?
Give Bees a Chance

Whether you've used lined paper tubes or routed tunnels, it's important to remove and clean all the cocoons in the fall. It's easy to do and doesn't hurt the bees one bit. In fact, chances are you will be saving their lives.

If there are pollen mites in each cell, the bees get covered in them, rendering them flightless and vulnerable to predators. Similarly, if there are fungal spores in the tubes or tunnels, they are spread by emerging bees.

Use your microscope or magnifying glass to take a close look at what's in your condo. A healthy cocoon will be surrounded by little pellets of bee poop, or "frass." These are usually black, but can vary with the colour of the pollen the larva ingested. You'll also see orange masses of mature hairy-footed pollen mites and little white dots that are immature stages of that mite. You may see larvae of predatory wasps. Look for any bee cocoons that are damaged and empty and put them in the garbage pile. If your tunnels are filled with crumbly brown matter and chalky white, hard, dead larvae, you may have a condo infected with a disease called "chalk brood." In this case, quarantine your condo and clean it with water and bleach. Sterilize your hands to prevent the disease from spreading.

For those cocoons that look viable, it's time to

In my backyard, the ground-nesting bees showed me their favourite patch of soil, so I added a few landmarks to help them orient themselves.

clean them. Soak the cocoons in a 5 percent solution of bleach in tepid water for 5 minutes. Wait for healthy cocoons to float to the top during that time. The cocoons are waterproof and the bleach will not affect the bees. Place the cocoons in a strainer and rinse gently in lukewarm water to dislodge mites and debris. Replace the solution with clean water and soak twice more, for five minutes each time, rinsing in between. Remove the cocoons and let them dry on a clean, dry towel. Cocoons that sink should be discarded.

Next, you can "candle" the bees by laying them on a powerful flashlight or light tray. Candling helps you see which cocoons contain bees and which may have been taken over by predators. If you don't see the shadow of the bee inside, it has likely been parasitized and you can either hatch it in a jar to see what emerges, or discard the cocoon. Keep track of how many large (female) cocoons and smaller (male) cocoons were intact.

To store the healthy cocoons, lay a wet piece of paper towel on the bottom of a glass or plastic container with small holes punched in the lid. Place the cleaned cocoons in an open plastic bag inside the container and set it in the fridge. You don't want the bees to dry out, nor do you want the cocoons to develop mould. Check your bees once a week and moisten or replace the paper towel with a drier one if needed. Look for mason bees in flowers as a sure sign that it's time to put your cocoons outside. Don't wait too long to release the cocoons—you don't want to chill the bees for more than 220 days.

A CALENDAR FOR MASON BEES

Female blue orchard mason bees live up to one month and produce one to six nest tunnels, with an average of five to fifteen eggs. Familiarize yourself with the seasonal tasks that will support these gentle bees in their quest to go forth and multiply. This calendar applies primarily to the Pacific Northwest; in colder zones, you will have to move back the release date by a month or two, depending on when the trees are in bloom.

Wild sunflowers are perfect for restoration projects where they can spread freely without crowding out less aggressive species.

February through March

- ✼ The final months for sourcing cocoons.

March

- ✼ In early March, reinstall routed nesting blocks that have been cleaned and disinfected with eco-friendly bleach. Or put out nesting blocks with new paper liners. Re-drilling old holes also helps clean them, but freshly drilled holes are best.
- ✼ Before releasing cocoons in late March or April, make sure there is a source of wet clay near the condo, or dig a wedge in the ground in a shady spot nearby. Ideally, the outdoor temperature should reach a high of 14C (57F) before you release the cocoons. This may be later in cooler zones.
- ✼ You have the option of releasing the cocoons in weekly increments (one-third at a time) to boost their chances in the event of inclement weather. About one-quarter of the blossoms in each cherry or apple tree should be in bloom to support the emerging bee.

April

- ✼ Adults are emerging, mating, foraging and laying eggs.
- ✼ Some females will return to their original nest; others will move on to nest in other locations.

May

- ✼ The egg hatches and the larva begins consuming pollen.
- ✼ Keep the condo in place until the fall. It's important to not jostle the immature bees in the condo tunnels until they develop into adults.

June

- ✼ By late June, the developed larva begins spinning a cocoon.

July

- ✼ The larva transforms into a white pupa inside the cocoon.

September

- ✼ The pupa makes its final transformation into an adult and waits for the warm spring weather before it emerges from the cocoon and then from the tunnel (if it is still inside one).
- ✼ Because they have adapted to the changeable temperatures in fall weather, the bees in the cocoons are not as sensitive to temperature changes now.

October

~ Time to clean the mason-bee cocoons and tunnels.

~ Place cocoons into cold storage and check weekly for mould or dryness. Wash off any mould, drying the cocoon thoroughly by placing it on an absorbent cotton towel, then return to the fridge with a dry paper towel. If cocoons are becoming too dry in the fridge, moisten a fresh piece of paper towel and return to the fridge.

bee byte

If you have sidewall shingles on your home, you may notice mason bees nesting in the gaps between them. These little bees will not harm your house and will pollinate the bee-friendly flowers in your spring garden.

The life cycle of a bee is connected to those of the plants that sustain it. Each species of bee will have indicator plants that signify the important milestones in its life. Bumblebee queens in the Pacifc Northwest, for example, emerge to pollinate salmonberry, red-flowering currant and Indian plum. In the east, queens begin feasting on Dutchman's breeches, Virginia bluebells and Atlantic camas. In the autumn, new queen bumblebees gorge on goldenrod and asters to build up their fat stores for hibernation. Mason bees hatch when the cherries are blooming. Spring mining bees emerge when hawthorn buds have opened.

MEADOW MATES: BLUE CAMAS, SEABLUSH AND NODDING ONION

In the spring on Vancouver Island, BC, blankets of colourful Garry oak meadow wildflowers create bee synergy, blooming in spring to help give nesting native bees a healthy boost. Take inspiration from this ecosystem and add this pollination powerhouse to your garden. Choose a sunny spot with excellent drainage, especially for blue camas (*Camassia quamash*), which blooms at the same time as seablush (*Plectritis congesta*). The bright-pink flowers of seablush have purple stamens producing a vibrant yellow pollen that, like blue camas pollen, provides mason bees and bumblebees with crucial nourishment for their broods. Add the pale-pink blossoms of nodding onion (*Allium cernuum*) for a slightly later-blooming follow-up feast of nectar and pollen for short- to long-tongued bees.

One of the best bulbs to plant for bees is the native blue camas, which provides important food for blue orchard mason bees and bumblebees.

Rocky Mountain bee plant (*Cleome serrulata*) is the correct cleome for bees. Plant this instead of the common garden cleome (*C. spinosa*).

ALPINE-MEADOW ALLIES: ROCKY MOUNTAIN AND YELLOW BEE PLANTS AND LUPIN

Create your own Rocky Mountain meadow by combining three bee plants so generous with their seeds that you can share them with the whole neighbourhood, making the hills come alive with the sound of bee symphonies. Much loved by bumblebees, Rocky Mountain bee plant (*Cleome serrulata*) is known by several common names, including "Navajo spinach" and "beeweed." The plant has a rich ethnobotanical history, with potential for use as a dye or paint plant. The long seed pods are spectacular, hanging from the flowers like spider legs. Another species of cleome, yellow bee plant (*Cleome lutea*) attracts more than 140 species of native bees while also feeding butterflies and wasps. (Even wasps play a beneficial role in keeping the garden in balance.) Complete the trio by adding wild lupin (*Lupinus perennis*), which provides a generous amount of pollen for bumblebee queens to feed to their spring brood.

Opposite, clockwise from top left: The Tennessee coneflower was once a very rare plant, but has been saved by gardeners who appreciate its delicate beauty. Long-headed coneflower is an emblem for prairie restoration for bees. I have liberated a few bees who got their legs caught in milkweed flowers. Plains coreopsis has rings visible to the human eye, as well as to bees.

Above: Wild bee balm (*Monarda fistulosa*) and fireweed (*Chamerion angustifolium*) are essential plants for ecological restoration. **Opposite, clockwise from top left:** A mining bee (*Andrena prunorum*) in rattlesnake master. A golden northern bumblebee (*Bombus fervidus*) in thistle. A green sweat bee (*Agapostemon*) in Bolander's phacelia. A cellophane bee (*Colletes*) in nodding onion.

Seeing the Prairies through Bee Goggles

Chris Helzer is an ecologist working for the Nature Conservancy who supervises the management and restoration of about 4,000 acres of land in Nebraska. He takes the life cycles of bees into consideration as he evaluates strategies to protect and restore native habitat. "One of the best outcomes from learning a lot about a group of species is that I start to see prairie through their eyes," he says. "That perspective has been really valuable for me and has led me to evaluate, restore and manage prairies differently . . . As a result, when I look at the prairies these days, I sometimes feel like I'm doing so through 'bee goggles.'"[33] Helzer notes that New England aster is one of the best native flowers for bees, attracting bumblebees, honeybees, large leafcutter bees and mining bees.

PRAIRIE SENTINELS: TALL, PURPLE AND LONG-HEADED CONEFLOWERS

Purple coneflower (*Echinacea angustifolia*) is a trifecta plant, providing nectar and pollen for short- and long-tongued bees, nectar for monarch butterflies and seeds for songbirds. Give a few stems a stubble cut for cavity-nesting bees after they bloom and begin to dry out in late summer, and it provides housing too. Long-headed coneflower (*Ratibida columnifera*), also billed as "Mexican hat," can be long-blooming, providing nectar and pollen from spring into fall. The most common form has yellow rays; a beautiful variation with burgundy rays is *R. columnifera* var. *pulcherrima*. Tall coneflower (*Ratibida pinnata*) is an excellent prairie meadow plant for bees and beneficial insects. Grow all three heights of prairie coneflowers in a perennial border with other prairie wildflowers such as aster, firewheel (*Gaillardia pulchella*) and goldenrod. This native-plant Victory Border will bloom in late summer when competition for forage is fierce and native bees really need your support.

AN EASTERN BOGGY TRIO: JOE-PYE, SWAMP MILKWEED AND BLUE VERVAIN

Joe-Pye weed (*Eutrochium purpureum*) is a bee-garden essential—bumblebees and honeybees scramble over its blossoms to collect pollen, and the dried hollow stems make good nesting sites. Add blue vervain to extend bloom into late summer and fall and to support even the smallest of bees. And make this a terrific trio of wildflowers that love moist soil and support butterflies with the addition of swamp milkweed (*Asclepias incarnata*), a surprisingly useful plant with fluffy floss attached to seeds that pop out of the beautiful boat-shaped pods. Reported to be six times more buoyant than cork, milkweed fluff was used during World War II to stuff military life jackets.[34]

A blond male bumblebee forages in nodding onion, an edible wildflower that attracts a number of native bee species.

NEW WORLD *natives*

A SEASONAL PLANTING CHART

*C*hoosing native bee plants will greatly help the survival of bees indigenous to your neighbourhood. Near-natives have also proven valuable to local pollinators. Please do not remove any plants from the wild; plants may be sourced instead from local native-plant clubs and botanical-garden and specialist nurseries.

PLANT ORIGIN

OW: Old World (indigenous to Europe, Asia or Africa)

NW: New World (indigenous to the Americas and their islands)

BLOOM PERIOD

E: Early

M: Mid-season

L: Late

SUC: May be seeded in succession

HARDINESS

refers to the coldest zone the plant is hardy to.

DEER AND DROUGHT RESISTANCE are most reliable once the plant is established.

BEES ATTRACTED

BB: Bumblebees

HB: Honeybees

SB: Solitary bees

BI: Beneficial insects

PLANT NAME AND NATIVE REGIONS	PLANT FAMILY AND BLOOM	HARDINESS	MAXIMUM HEIGHT AND GARDEN NOTES	BENEFITS TO BEES*
Aster, Heath *Symphyotrichum ericoides* Most of Canada and US	M–L Aster. White daisy-like flowers with yellow centres.	**Perennial** Hardy to zone 3. Drought resistant.	**2 ft (60 cm)** Sun, average to dry soil. Seeds need light to germinate. Excellent blossom density for bees.	**BB, HB, SB, BI** Honey plant. Cuckoo, mining, plasterer, small carpenter, sweat bees.
Aster, New England *Symphyotrichum novae-angliae* SK, MB, OT, QC, NB, NS, eastern US	M–L Aster. Daisy-like flowers with yellow centres and pink, mauve or white petals. Can spread aggressively.	**Long-blooming Perennial** Hardy to zone 3. Deer and drought resistant.	**40–80 in (1–2 m)** Sun, well-drained soil. Seeds need cold, moist stratification unless planted in fall. Pinch back May or June. Divide biennially.	**BB, HB, SB, BI** Honey potential for *Symphotrichum* spp.: 30–50 lbs/ac. Fall source of pollen and nectar. Cuckoo, green sweat, leafcutter, long-horned, mining, small carpenter bees.
Baby Blue Eyes *Nemophila menziesii* AK, CA, NV, OR	E–M–SUC Borage. Shallow light-blue bowl, white centre. *N. menziesii* var. *atomaria* has speckled nectar guide.	**Self-seeding Annual**	**1 ft (30 cm)** Sun to dappled shade, soil amended with compost and sandy soil. Direct-sow early spring or fall. Light aids germination. Deadhead. Suitable for pots.	**BB, HB, SB, BI** Nectar and pollen easily accessed by wide variety of bees. Key shade plants for bees. Specialist pollinator: *Dufourea nemophilae*.
Bee Plant, Rocky Mountain *Cleome serrulata* BC, Prairies, ON, QC, US Great Plains, Rockies	M–L Caper. Purple-pink and white flowers with long, spidery stamens. Long seed pods similar to kale.	**Self-seeding Annual** Deer and drought resistant.	**1–5 ft (30–150 cm)** Sun, well-drained soil. Sow as a wildflower in early spring or late fall. Unpalatable for livestock.	**BB, HB, SB** Honey plant. Long-tongued bees sip nectar; small bees perform acrobatics to gather pollen.
Bee Plant, Yellow, a.k.a. Spider Plant *Cleome lutea* Western US	M–L Caper. Yellow flowers with long, spidery stamens. See Rocky Mountain bee plant. Can be invasive.	**Self-seeding Annual** Deer and drought resistant.	**1–5 ft (30–150 cm)** Seeds may be difficult to obtain. Unpalatable for livestock. Yellow spider plant can be weedy.	**BB, HB, SB** Sugar concentration in *C. lutea* is up to 23 percent. Special value to native bees. See Rocky Mountain bee plant.
Blanket Flower *Gaillardia aristata* YT, BC, Prairies, ON, QC, northern and western US	M–L Aster. Daisy-like with yellow, orange and red petals, red-brown centre.	**Perennial** Hardy to zone 2. Deer and drought resistant.	**1–3 ft (30–90 cm)** Sun, dry to medium, well drained loam. Seeds need light to germinate. Plant early spring. Deadhead. Cut back in fall. Divide in 2–3 years.	**BB, HB, SB, BI** Honey plant. Sugar concentration 32 percent. Leafcutter, long-horned, small carpenter, sweat bees.
Blazing Star, Dotted *Liatris punctata* Central North America	M–L Aster. Upright wands of fringed mauve flowers. **Group with:** Drumstick allium.	**Long-blooming Perennial** Hardy to zone 3.	**2 ft (60 cm)** Sun to partial shade in medium to dry conditions. Seeds benefit from cold, moist stratification.	**BB, SB, BI** Mining, large leafcutter, small carpenter bees, monarch butterfly. Large swaths really attract bees.
Blazing Star, Marsh *Liatris spicata* ON, QC, eastern US	L Aster. Cultivars in white, pink, purple. See dotted blazing star.	**Perennial** Hardy to zone 4.	**2–4 ft (60–120 cm)** Sun to partial shade in mix of sand, loam and clay, or soil high in clay. See dotted blazing star.	**BB, SB, BI** Late-blooming honey plant. Plant with boneset, swamp milkweed. See dotted blazing star.
Bluebells, Virginia *Mertensia virginica* ON, QC, eastern US	E Borage. Bell-shaped flower, long tubular corolla changes from pink to blue.	**Perennial** Hardy to zone 3. Deer resistant.	**1–2 ft (30–60 cm)** Dappled shade, average to moist, fertile, well-drained soil. Sow seed in fall (needs cold stratification) or divide.	**BB, HB, SB** Long-tongued, mason, mining bees.

*Each pound per acre is equal to approximately 1.1 kilograms per hectare.

PLANT NAME AND NATIVE REGIONS	PLANT FAMILY AND BLOOM	HARDINESS	MAXIMUM HEIGHT AND GARDEN NOTES	BENEFITS TO BEES*
Blue-eyed Grass, a.k.a. Grasswidow *Sisyrinchium montanum* Canada, most of US	M Iris. See Idaho blue-eyed grass. **Group with:** Other *Sisyrinchium* spp.	Perennial Hardy to zone 3.	6–12 in (15–30 cm) Self-seeding. Can be mowed after blooming. Suitable for pots, roof-top garden, pond edge. See Idaho blue-eyed grass.	SB, BI Short-tongued bees such as sweat and masked.
Blue-eyed Grass, Idaho *Sisyrinchium idahoense* BC, western US	M Iris. Dainty periwinkle-blue flower, yellow centre, clear nectar guides. See blue-eyed grass.	Perennial Hardy to zone 4.	4–15 in (10–38 cm) Sun in moist, well-drained, average to fertile soil with high lime. Sow seeds indoors or in cold frame in spring or direct early fall. See blue-eyed grass.	SB, BI Short-tongued bees such as sweat and masked. Support small bees with plantings in pots, roof-top gardens, on pond edges.
Blue-eyed Mary, a.k.a. Giant Collinsia, Blue Lips *Collinsia grandiflora* Pacific Northwest	E–M Plantain. Tiers of double-lipped blossoms. Top lips lighter lavender than bottom. See spring blue-eyed Mary.	Self-seeding Annual	4–15 in (10–38 cm) Important native-bee shade plant. Seed requires light for germination. Press seed into soil in late fall. Thin to 1 ft (30 cm) apart. Water regularly. See spring blue-eyed Mary.	BB, SB, BI Bumblebees and mason bees. Large bees trigger lips, dust-ing their bodies with pollen. Butterfly plant. Conifer woodland bee plant.
Blue-eyed Mary, Spring *Collinsia verna* ON (at risk), eastern North America	E–M Plantain. **Group with:** Bolander's phacelia, Chinese houses. See blue-eyed Mary.	Self-seeding Annual	4–15 in (10–38 cm) Plant varying *Collinsia* spp. for native bees. See blue-eyed Mary.	BB, SB, BI Long-horned, mason, small car-penter bees. Essential woodland bee plant. See blue-eyed Mary.
Brown-eyed Susan *Rudbeckia hirta* SK, ON, QC, eastern and central US	M–L Aster. Daisy-like with yel-low petals, brown centre. **Group with:** Goldenrod, Joe-Pye weed, purple coneflower, scarlet bee balm, tall coneflower.	Biennial or Short-lived Perennial Hardy to zone 3. Deer and drought resistant.	2–5 ft (60–150 cm) Grows in sun or partial shade in medium-fertile, well-drained soil. Seed needs cold stratification, or start in fall. Self-seeding. Divide in spring or fall. Selectively cut flowers to encourage blooms.	BB, HB, SB, BI Cuckoo, green sweat, leafcutter, long-horned, small carpenter, sweat bees. Specialist pollina-tors: *Andrena rudbeckia* and *Heterosarus rudbeckia*. Attracts bees in long-blooming patches.
California Bluebells *Phacelia campanularia* CA	E–M Borage. Cup-shaped vibrant blue blossom. **Group with:** Calendula, poached egg flower.	Annual Drought resistant.	8–20 in (20–50 cm) Sunny, well-drained soil. Direct-sow early spring or fall. Suitable for pots; interplant with vegetables.	BB, HB, SB, BI Especially attractive to mason and mining bees. Water in dry season to increase nectar production.
Camas, Blue *Camassia quamash* Pacific Northwest	E Asparagus. Striking blue-purple flower, protruding stamens. Plant in wet meadow that dries in summer.	Perennial Hardy to zone 3.	1 ft (30 cm) Sun to light shade. Moist, heavy soil. Plant bulbs in late fall 4 in (10 cm) deep, 6 in (15 cm) apart. Let foliage die back before trimming.	BB, SB Nectar and yellow to violet pollen. Important early source of pollen for queen bumblebees.
Chinese Houses *Collinsia heterophylla* BC, western US	E–M Plantain. Tiers of double-lipped blossoms. Top lips white; bottom purple, pink or blue.	Self-seeding Annual	20 in (50 cm) Shade; rich, moist soil. Start indoors early spring or direct-sow spring or fall. Seeds need light to germinate. Grows well under deciduous trees.	BB, SB Key shade plant for native bees such as bumblebees and mason bees. Large bees trigger lips, dusting their bodies with pollen.
Columbine, Wild *Aquilegia canadensis* eastern North America	E–M Buttercup. Showy red and yellow flower, protruding stamens. Deep corollas resemble talons.	Perennial Hardy to zone 3. Deer resistant.	1–2 ft (30–60 cm) Sun or shade, dry to medium soil. Seeds need light. Taproot; transplant small. Indicator of return of migrating hummingbirds.	BB, SB Early-spring nectar for bumble-bee queens. Small bees also access nectar and pollen with acrobatics.

PLANT NAME AND NATIVE REGIONS	PLANT FAMILY AND BLOOM	HARDINESS	MAXIMUM HEIGHT AND GARDEN NOTES	BENEFITS TO BEES*
Coneflower, Long-headed, a.k.a. Mexican Hat *Ratibida columnifera* BC to QC, much of continental US	M–L Aster. Elongated head with small florets, yellow or burgundy drooping rays.	Perennial Hardy to zone 2. Deer resistant.	**1–2 ft (30–60 cm)** Sun to partial shade, well-drained soil in medium-dry to dry conditions. Seeds need cold stratification or plant in fall. Suitable for pots.	BB, HB, SB, BI Cuckoo, leafcutter, long-horned, sweat bees. Special value to native bees. Visited for nectar from many insects, including butterflies and moths.
Coneflower, Purple *Echinacea angustifolia* SK, MB, central US	M–L Aster. Purple petaloid rays, rust-orange centre. Cultivars in variety of colours. **Group with:** Anise hyssop, blazing star, brown-eyed Susan.	Long-blooming Perennial Hardy to zone 3.	**12–18 in (30–46 cm)** Sun in a variety of soils, moderate moisture. Seeds do better with cold stratification. Prepare bed by digging compost into well-loosened soil, add compost each spring. Suitable for pots.	BB, HB, SB Digger, green sweat, leafcutter and sweat bees. Specialist pollinator: mining bee *Andrena helianthiformis*. Monarch, swallowtail butterflies.
Coneflower, Tall *Rudbeckia laciniata* BC, MB, ON, QC; central, eastern US	M–L Aster. Daisy-like with yellow petals, light-green centre. **Group with:** Boneset, swamp milkweed.	Perennial Hardy to zone 3.	**10 ft (3 m)** Sun to shade in medium to moist conditions. Seed requires cold, moist stratification unless planted in the fall.	BB, HB, SB, BI Honey plant. Cuckoo, leafcutter, long-horned, mining, small carpenter, sunflower, sweat bees. Specialist pollinator: *Heterosarus rudbeckia*.
Coneflower, Yellow, a.k.a. Grey-headed Coneflower *Ratibida pinnata* QC, eastern US	M–L Aster. Grey head with small brown florets, long drooping yellow rays. Anise scent when crushed.	Perennial Hardy to zone 3. Deer resistant.	**4 ft (1.2 m)** Sun to partial shade, well-drained soil. Seeds need cold stratification. Sow indoors or in cold frame during winter, early spring or fall. Divide in spring.	BB, HB, SB, BI Cuckoo, leafcutter, long-horned, sweat bees. Specialist pollinator: *Andrena rudbeckia*. Water to stimulate nectar.
Dutchman's Breeches *Dicentra cucullaria* MB, ON, QC, NB, NS, PEI, WA, OR, ID, eastern US	E Poppy. Ivory and white flower looks like tiny upside-down trousers. Toxic. **Group with:** Hardy geranium, Solomon's seal.	Perennial Hardy to zone 3. Deer resistant. .	**1 ft (30 cm)** Dappled shade, fertile moist soil. Seeds require warm, moist stratification 60–90 days, followed by cool, moist stratification 60–90 days. Or plant bare-root stock spring or fall.	BB, HB, SB, BI Mason, mining bees. Early-spring plant for long-tongued bees. Look for signs of nectar robbing. Woodland bee plant.
Firewheel *Gaillardia pulchella* MB, ON, QC, eastern US	M–L Aster. Daisy-like with yellow-tipped red petals, red centre. **Group with:** Long-headed coneflower, prairie smoke.	Perennial Hardy to zone 3. Deer and drought resistant.	**1–2 ft (30–60 cm)** Sun, dry to medium, well-drained loam. Seeds need light. Plant early spring. Deadhead. Cut to 6 in (15 cm) in fall. Divide in 2–3 years.	BB, HB, SB, BI Long-blooming honey plant. Sugar concentration 32 percent. Leafcutter, long-horned, small carpenter, sweat bees.
Five Spot *Nemophila maculata* CA	E–M–SUC Borage. Deep-purple spot on each of 5 white petals.	Self-seeding Annual	**6 in (15 cm)** Sun to dappled shade, sandy soil amended with compost. See baby blue eyes.	BB, HB, SB, BI Special value to native bees. See baby blue eyes.
Flax, Western Blue *Linum lewisii* Northwest, central North America	E–M Flax. Periwinkle-blue flower with darker nectar guides.	Short-lived Perennial Hardy to zone 4.	**1 ft (30 cm)** Sunny, dry-medium soil, sand, loam. Self-seeding. Suitable for pots, between vegetables.	HB, SB, BI Leafcutter, mining, sweat bees. Special value to native bees for nectar.
Garlic, Wild *Allium canadense* Central, eastern US	E Amaryllis. Gorgeous globes of white to pink florets.	Perennial Hardy to zone 4. Deer resistant.	**1 ft (30 cm)** Sun to part shade, dry to medium sand to loam. Sow from fresh seed left outside to stratify. Self-seeding.	BB, HB, SB, BI Mason, plasterer, sweat, masked bees. Specialist pollinator: onion bee, *Heriades carinatum*.

PLANT NAME AND NATIVE REGIONS	PLANT FAMILY AND BLOOM	HARDINESS	MAXIMUM HEIGHT AND GARDEN NOTES	BENEFITS TO BEES*
Geranium, Sticky *Geranium viscosissimum* Pacific coast from BC to CA inland from SK to WY and SD.	**E–M–L** Geranium. Shallow pink or pink/purple flower with darker nectar guides. Some varieties spread aggressively. **Group with:** Columbine, Oregon tea.	**Perennial** Hardy to zone 4. Deer and drought resistant.	**1 ft (30 cm)** Prefers dappled shade and moist, rich, well-drained soil, but adaptable. Propagate by seed or division. Soak seed 48 hours before planting.	**BB, HB, SB, BI** Attracts cuckoo, long-horned, mason, mining, small carpenter, sweat bees. Grow for them in shallow soil under trees and eaves, difficult-to-plant areas.
Geranium, Wild *Geranium maculatum* MB, ON, QC, eastern US above FL	**E–M–L** Geranium. See sticky geranium. **Group with:** Solomon's seal.	**Perennial** Hardy to zone 4. Deer and drought resistant.	**1 ft (30 cm)** Very easy to propagate and super-hardy in dry shade, although more lush in moister settings. See sticky geranium.	**BB, HB, SB, BI** Specialist pollinator: *Andrena distans*. See sticky geranium.
Germander, Canada *Teucrium canadense* North America, excluding far north	**M–L** Mint. Spikes of white to lavender flowers, purple nectar guides. Can spread aggressively. **Group with:** Anise hyssop.	**Perennial** Hardy to zone 4. Deer resistant.	**1–3 ft (30–90 cm)** Sun to dappled shade, moist soil. Seeds require cold, moist stratification. Pond edges.	**BB, HB, SB, BI** Nectar. Long-tongued bees such as bumblebees, cuckoo, leaf-cutter, mining bees. Honey plant. Butterflies, hummingbirds.
Gilia, Bird's Eye *Gilia tricolour* CA	**M–L** Phlox. White blossoms with dark-violet eye, chocolate scent. Plant in pots.	**Self-seeding Annual**	**2 ft (60 cm)** Sun, loam to sandy well-drained soil, dry conditions. Sow indoors 4 weeks before last frost. Press seed into soil.	**BB, HB, SB** Nectar and bright-blue pollen. Attracts bees of all stripes, hummingbirds. In California, mason bees.
Gilia, Globe, a.k.a. Blue Thimble Flower, Queen Anne's Thimble *Gilia capitata* US Pacific coast	**M** Phlox. Globe-shaped clusters of 50–100 tiny blue flowers. **Group with:** *Gilia* spp. and *Nemophila* spp.	**Self-seeding Annual**	**1–3 ft (30–90 cm)** Sun to part shade, clay to loam in hot, dry conditions. Sow indoors 8 weeks before last frost. Press seed into soil. Plant out when soil warms.	**BB, HB, SB** Blue pollen. Green sweat, mining bees, more. Keystone plant for native pollinators. Deadhead and water to encourage nectar.
Golden Alexander, a.k.a. Meadow Parsnip *Zizia aurea* SK, eastern Canada, eastern US	**E–M** Carrot. Umbels of yellow-green flowers. **Group with:** Columbine, false blue indigo, lupin, perennial sage.	**Perennial** Hardy to zone 3.	**12–30 in (30–76 cm)** Sun to part shade. Loamy soil, can tolerate rockiness in medium to moist conditions. Can plant in fall but tricky to germinate; instead, let it readily self-seed.	**BB, SB, BI** Cuckoo, green metallic, mason, masked, mining, small carpenter, sweat bees. *Andrena ziziae* is oligolectic. In some regions, parallels mason bee nesting.
Goldenrod, Canada *Solidago canadensis* Northeastern North America	**M–L** Aster. Spikes of small yellow flowers. Can spread aggressively. **Group with:** Anise hyssop, aster, sedum, purple coneflower, sunflower.	**Perennial** Hardy to zone 0. Deer resistant.	**2–5 ft (60–150 cm)** Sun to partial shade in medium, well-drained soil. Dry to moderate conditions. Cold stratify seeds or plant in fall, barely covering. Suitable for pond edges. Edible plant; contraindications: pregnancy, heart and kidney concerns.	**BB, HB, SB, BI** Honey potential (*Solidago* spp.): 25–50 lbs/ac. Honey is dark with a strong flavour, rich in protein, minerals. Key source of late-season pollen for long- and short-tongued bees. Large carpenter, mining, sweat bees.
Goldenrod, Early *Solidago juncea* Eastern Canada and US	**M–L** Aster. Blooms earlier than other goldenrods. Can spread aggressively. **Group with:** Tall coneflower.	**Perennial** Hardy to zone 3. Deer resistant.	**2–4 ft (60–120 cm)** Sun to part shade in medium, dry to moderate well-drained soil. Cold stratify seed or plant in fall, barely covering.	**BB, HB, SB, BI** Grow extra goldenrod plants in pots to support bees of all stripes. See Canada goldenrod.

Male cactus bees (*Diadasia*) sleep inside flowers and chase off other males from their chosen mating site. ANNA HOWELL PHOTO

PLANT NAME AND NATIVE REGIONS	PLANT FAMILY AND BLOOM	HARDINESS	MAXIMUM HEIGHT AND GARDEN NOTES	BENEFITS TO BEES*
Harebell *Campanula rotundifolia* Canada, temperate US	M–L Bellflower. Light-purple bell-shaped flower, delicate stem. Can spread aggressively. **Group with:** Prairie smoke, western blue flax.	**Perennial** Hardy to zone 3.	**6–20 in (15–50 cm)** Sun or dappled shade, sand to loam, moist to dry conditions. Tolerates alkaline soil. Suitable for pots.	**BB, HB, SB** Honey plant. Leafcutter, long-horned, mason, mining, small carpenter, sweat bees. Butterfly and hummingbird plant.
Heather, Pink Mountain *Phyllodoce empetriformis* YT, NU, BC, AB, northwest US	E–M Heather. Delicate pink bell-shaped flower. Native alternative to Scottish heather.	**Perennial** Hardy to zone 4.	**15 in (38 cm)** Found in alpine, subalpine rocky habitat, sun to partial shade. Acidic, well-drained soils. Propagate from cuttings.	**BB** Important plant for the conservation of bumblebee species found in alpine habitat. Grows in pots, too.
Hyssop, Giant Yellow *Agastache nepetoides* Eastern North America	M–L Mint. Tall towers of cream-coloured lipped flowers. Can be weedy. **Group with:** Goldenrod; plant below maple trees.	**Perennial** Hardy to zone 2. Deer and drought resistant.	**6 ft (1.8 m)** Partial sun to light shade; moderately wet to somewhat dry conditions in fertile, loamy soil. Seeds need cold, moist stratification and light to germinate, or plant in fall.	**BB, HB, SB, BI** Nectar, some pollen. Sweat, masked bees. Honey plant. Special value to native bees. Significant bumblebee plant.
Indigo, False Blue *Baptisia australis* QC, central and eastern US	E–M Legume. Blue-purple pea-shaped flower. Toxic. **Group with:** White or purple prairie clover, lupin.	**Perennial** Hardy to zone 3. Drought resistant.	**3–4 ft (90–120 cm)** Sun, with some shade tolerance. Average to poor, sandy, well-drained soil. Scarify seeds with sandpaper, apply inoculant. See wild indigo.	**BB, HB, SB** Long-horned bees. Loved by long-tongued bumblebees and their queens for pollen and nectar rewards. Helps fill June nectar gap. Monarch butterfly plant.
Indigo, Wild *Baptisia tinctoria* ON, eastern US	E–M Legume. Yellow to cream, or white (*B. alba*). See false blue indigo.	**Perennial** Hardy to zone 4. Drought resistant.	**2–4 ft (60–120 cm)** See false blue indigo. Sow indoors early spring or direct-sow fall. Trim frost-damaged foliage spring.	**BB, HB, SB** See false blue indigo. Special value to native bees, especially bumblebees.
Ironweed, Prairie *Vernonia fasciculata* SK, MB, ON, central US	M–L Aster. Purple-fringed blossom for late-season colour. **Group with:** Boneset, swamp milkweed.	**Perennial** Hardy to zone 3.	**2–6 ft (60–180 cm)** Sun to part shade; medium-moist sand, loam, clay. Seeds need cold, moist stratification or plant in fall. Self-seeding. Divide in early spring.	**BB, HB, SB, BI** Fall honey plant. Long-tongued cuckoo, large leafcutter and mining bees. Monarch butterfly nectar plant.
Joe-Pye Weed *Eutrochium purpureum* Eastern and central North America	M–L Aster. Loose clusters of fringed pink flowers. See spotted Joe-Pye weed.	**Perennial** Hardy to zone 3. Deer resistant.	**40–80 in (1–2 m)** Sun to part shade in sand, loam or clay. Medium to moist. Seeds need light to germinate. Divide plants in spring or fall. See spotted Joe-Pye weed.	**BB, HB, SB, BI** Fall honey plant. Leafcutter, mining, small carpenter, sweat bees. Significant plant for native bee conservation.
Joe-Pye Weed, Spotted *Eutrochium maculatum* ON, QC to KY	M–L Aster. **Group with:** bee balm, New England aster. See Joe-Pye weed.	**Perennial** Hardy to zone 4. Deer resistant.	**2–6 ft (60–180 cm)** Suitable for pots, rooftop, pond edge. Stubble cut in fall. Hollow stems are great for bee hotels. See Joe-Pye weed.	**BB, HB, SB, BI** See Joe-Pye weed. Significant butterfly and moth plant. Water during drought to ensure nectar production.
Larkspur, Blue Wild *Delphinium carolinianum* MB and central and southern US	E–M Buttercup. Blue to violet flower with long nectar tube. Toxic. **Group with:** *Gaillardia* spp., long-headed coneflower.	**Perennial** Hardy to zone 3. Deer resistant.	**4 ft (1.2 m)** Sun to part shade, in well-drained, sandy soil, dry to moderate conditions. Seeds need cold, moist stratification, or plant in fall.	**BB** Long-tongued bumblebees. Plant various wild larkspurs to extend bloom for bees, but avoid cultivars because they lack rewards for bees.

PLANT NAME AND NATIVE REGIONS	PLANT FAMILY AND BLOOM	HARDINESS	MAXIMUM HEIGHT AND GARDEN NOTES	BENEFITS TO BEES*
Lupin, Wild *Lupinus perennis* North America	**E–M** Legume. Glorious towers of purple-lipped blossoms. **Group with:** False blue indigo, yarrow.	**Perennial** Hardy to zone 3. Deer resistant.	2 ft (60 cm) Sun, dry, well-drained, sandy soil. Direct-sow after danger of frost: scarify seed with sandpaper, then provide cold, moist stratification or plant in fall. Treat with inoculum before sowing. Self-seeding.	**BB, SB** Mason and sweat bees. Pollen is ejected onto the visiting pollinator's head. Essential native bee conservation plant, especially for bumblebees. Butterfly plant.
Milkweed, Butterfly *Asclepias tuberosa* ON, QC, CA, central and eastern US	**M** Dogbane. Clusters of bright-orange flowers. The pollen is made up of chains, rather than grains. Toxic.	**Long-blooming Perennial** Hardy to zone 4. Deer and drought resistant.	2–3 ft (60–90 cm) Sun to part shade, sandy to loam, dry to medium soil. Easy to start from seed; direct-sow. Deep taproot, does not like to be moved. Divide early spring or fall.	**BB, HB, SB, BI** Honey potential: 120–250 lbs/ac, depending on soil quality. Blooms 8–10 weeks for cuckoo, leafcutter, mining, small carpenter, small resin, sweat bees.
Milkweed, Swamp, a.k.a. Red Milkweed *Asclepias incarnata* SK, MB, ON, QC, US east of CA	**M** Dogbane. Vanilla-scented pink flowers. **Group with:** Blue vervain, white turtlehead. Toxic. Endangered in some areas.	**Perennial** Hardy to zone 3. Deer resistant.	8–20 in (20–50 cm) Sun, medium to wet soil with clay or sand and good drainage. Seed needs cold, moist stratification or plant in fall. Pond edges.	**BB, HB, SB, BI** Milkweeds are important food for caterpillars of monarch butterfly. Many insects feed on it. Hummingbird plant. See butterfly milkweed for bee information.
Mint, Downy Wood *Blephilia ciliata* ON, Great Lakes, mid-Atlantic US	**E–M** Mint. Spears of bilabial mauve flowers. Endangered in Canada and northeastern US. **Group with:** Chinese houses, spotted bee balm.	**Perennial** Hardy to zone 4. Deer resistant.	1–2 ft (30–60 cm) Sun to part shade, in well-drained soil. Seeds need light to germinate and require cold, moist stratification. Spreads by rhizomes.	**BB, HB, SB** Leafcutter bees, small carpenter bees, masked bees and more.
Mint, Mountain *Pycnanthemum pilosum* ON, QC, eastern US south of Great Lakes, north of FL	**M–L** Mint. Clustered small, white double-lipped blooms; tiny purple dots are nectar guides for these open-access flowers that only open a few at a time.	**Perennial** Hardy to zone 3. Deer resistant.	1–3 ft (30–90 cm) Sun to part shade. Not fussy about soil, but prefers moist conditions. Direct-sow seeds in spring or fall; germination can take 30 days. Or divide plants in spring. Spreads by rhizomes, but not aggressive. Suitable for pond edges and pots.	**BB, HB, SB, BI** Very popular northwest bee plant. Cuckoo, green sweat, long-horned, resin, sweat and masked bees sip nectar at the mountain-mint buffet with shiny turquoise cuckoo wasps, butterflies and ants.
Onion, Nodding, a.k.a. Lady's Leek *Allium cernuum* BC, AB, SK, MB, ON, much of continental US	**M–L** Amaryllis. White to pink bell-shaped flowers nod at end of globular umbels. Edible stems, buds, flowers.	**Perennial** Hardy to zone 2. Deer and drought resistant.	1–2 ft (30–60 cm) Sun to part shade in sand to loam, dry to medium. Easy to grow from fresh seed collected and left in cold to stratify. See pink onion. Edible.	**BB, HB, SB, BI** Sweat bees and syrphid flies frequent visitors. Leafcutter, mining, small resin and sweat bees. Essential bee garden plant. See pink onion.
Onion, Pink *Allium stellatum* SK, MB, ON, central US	**M–L** Amaryllis. Blooms loved by bees of all stripes. See nodding onion.	**Perennial** Hardy to zone 2. Deer and drought resistant.	1 ft (30 cm) Self-seeding. Suitable for pots, around vegetables. See nodding onion.	**BB, HB, SB, BI** Heritage bee plants that once blanketed the landscape. See nodding onion.
Oregon Tea *Clinopodium douglasii* Pacific coast from AK to CA	**M** Mint. Small white bilabial flower. Minty scent. **Group with:** Grow under oak and arbutus trees.	**Perennial** Hardy to zone 5. Deer and drought resistant.	6 in (15 cm) Sun to dappled shade in moist soil. Seeds require cold, moist stratification, or plant in fall. Propagate by stem cuttings. Spreads to 80 in (2 m).	**BB, HB, SB** Long-tongued bees. Important shade plant for bees; grow with kinnikinnick and wild strawberry for a pollinator party.

PLANT NAME AND NATIVE REGIONS	PLANT FAMILY AND BLOOM	HARDINESS	MAXIMUM HEIGHT AND GARDEN NOTES	BENEFITS TO BEES*
Pearly Everlasting *Anaphalis margaritacea* Canada except NU; US except ND, southeast states	M–L Aster. Clusters of papery white blossoms with velvety gold centre. **Group with:** Aster.	Perennial Hardy to zone 3. Drought resistant.	1–3 ft (30–90 cm) Sunny site, not fussy about soil. Sow seeds in fall in outdoor pots. Young leaves can be eaten cooked. Dye plant. Can be a garden thug.	BB, HB, SB, BI Honey potential: 20 lbs (9 kg) per hive per season. Significant butterfly plant. Important medicinal plant for native bee conservation.
Penstemon, Foxglove *Penstemon digitalis* ON, QC, eastern US north of FL	E–M Plantain. White bilabial flowers with long corollas. **Group with:** *Campanula* spp., perennial sage.	Perennial Hardy to zone 2. Deer and drought resistant.	2–4 ft (60–120 cm) Sun to partial shade in moist to dry soil. Easy to start from seeds. Seeds require cold, moist stratification and light to germinate. Divide in spring or fall.	BB, HB, SB, BI Digger, mason, mining, large leafcutter, small carpenter, sweat and wool carder bees. Specialist pollinator: *Osmia distincta*.
Penstemon, Hairy *Penstemon hirsutus* ON, QC, northeastern US	E–M Plantain. Purple and white. *Penstemon* draws hummingbirds. See foxglove penstemon.	Perennial Hardy to zone 3. Deer and drought resistant.	20 in (50 cm) Plant instead of invasive foxglove. Long-tongued bees find nectar and pollen; small bees find pollen. See foxglove penstemon.	BB, HB, SB, BI Digger bees line their ground tunnel nests with oils secreted by the plants. See foxglove penstemon.
Penstemon, Rocky Mountain *Penstemon strictus* Southwestern US	E–M Plantain. Purple-blue flowers with violet nectar guides.	Perennial Hardy to zone 4. Deer and drought resistant.	3 ft (90 cm) See hairy penstemon. Useful for stabilizing soil.	BB, HB, SB, BI Significant bumblebee and hummingbird plant. See hairy penstemon.
Phacelia, Bolander's *Phacelia bolanderi* Pacific coast	E–M Borage. Shallow cup-shaped lavender flowers. **Group with:** Grow under oak trees.	Perennial Hardy to zone 8. Deer resistant.	1–2 ft 30–60 Sun with water or dry shade. Cover seed lightly in spring. Short-lived, but self-seeding. Suitable for pots.	BB, HB, SB Larger bees sip the nectar and smaller bees perform acrobatics to collect pollen.
Poppy, California *Eschscholzia californica* BC, AB, MB, ON, NS, much of continental US	M–L Poppy. Flowers have four petals, usually orange. Cultivars come in shimmering pastel shades. **Group with:** Lacey phacelia, poached egg flower, tidy tips.	Self-seeding Annual Deer and drought resistant.	1–2 ft (30–60 cm) Sun to dappled shade, well-drained, gritty soil. Light is required for germination. Sow directly into the ground in fall or early spring. Self-seeding. Mature pods pop open and fling tiny seeds. Deadhead to encourage blooms.	BB, HB, SB, BI Pollen. Sweat bees and other solitary bees visit to gather pollen, but bumblebees are most fond of this plant.
Prairie Crocus *Anemone patens*, *Pulsatilla patens* Canada west of QC, central US to TX	E Buttercup. Petals are blue fuzzy sepals; centres yellow with multiple stamens. Foliage can irritate skin. **Group with:** Allium.	Perennial Hardy to zone 3.	8 in (20 cm) Sun to part shade in sand to loam, dry to medium conditions. Stratify seeds. Feathery seed heads are beautiful in dried bouquets. Suitable for pots.	BB, HB, SB Abundant pollen, small nectar rewards. Mining, sweat bees. Key source of early-spring pollen for bumblebee brood.
Prairie Smoke *Geum triflorum* Northern North America	E–M Rose. White to pale-red petals hidden by showy red sepals and bracts. Feathery seed heads. **Group with:** Blanket flower, prairie crocus.	Perennial Hardy to zone 1.	1 ft (30 cm) Sun to part shade in well-drained sandy or rocky soil in dry conditions. Seeds require cold, moist stratification. Tricky to germinate; buy plants if you can. Suitable for pots.	BB Morphology of the flowers is bumblebee-specific, as they are strong enough to force the petals open. Look for nectar robbing. Small bees perform acrobatics to feed on pollen. Sweat bees.
Prickly Pear, Western *Opuntia polyacantha* BC, AB, SK; western, central US	E Cactus. Shallow waxy yellow blooms with multiple yellow stamens. **Group with:** Sedum.	Perennial Hardy to zone 4. Deer and drought resistant.	4–15 in (10–38 cm) Full sun for at least 6 hours per day, extremely well-drained poor soil. Suitable for pots. Can become weedy in overgrazed pasture.	BB, HB, SB Large carpenter, leafcutter, long-horned, mining, plasterer, sweat bees. Specialist pollinator: *Ashmeadiella opuntiae*.

PLANT NAME AND NATIVE REGIONS	PLANT FAMILY AND BLOOM	HARDINESS	MAXIMUM HEIGHT AND GARDEN NOTES	BENEFITS TO BEES*
Pussytoes, Field *Antennaria neglecta* Canada apart from NU, YT, NL; central and northeast US	E–M Aster. Fuzzy white and pink flowers can be wind- or insect-pollinated. See plantain-leaved pussytoes.	Perennial Hardy to zone 4. Deer resistant.	15 in (38 cm) Sun to light shade in poor soil, moist to medium conditions. Seeds need cold, moist stratification and light to germinate. Spreads by rhizomes. See plantain-leaved pussytoes.	HB, SB Cuckoo, mining, sweat bees. Many cultivars of this nectar- and pollen-rich plant are available to support native bees. See plantain-leaved pussytoes.
Pussytoes, Plantain-Leaved, *Antennaria plantaginifolia* QC, NS, eastern US	E–M Aster. See field pussytoes. **Group with:** Milkweed and low-growing sedum.	Perennial Hardy to zone 3.	6 in (15 cm) Thrives in poor soil. Does not compete well with other species. See field pussytoes.	HB, SB Bees seek nectar. Beetles feed on pollen. Butterfly plant. See field pussytoes.
Rattlesnake Master *Eryngium yuccifolium* Central and eastern US	M Carrot. Globes of milky white blossoms on long stems. Green anthers turn white or pink before releasing pollen. Prickly.	Perennial Hardy to zone 3. Deer and drought resistant.	1–4 ft (30–120 cm) Sun, thrives in poor soil; tolerates clay, doesn't like waterlogged soil. Easy to grow from seed with cold, moist stratification. Plants do not like to be disturbed.	BB, HB, SB, BI Masked, mining, bees. Stems suitable for tunnel-nesting bees. Important plant for beneficial insects.
Sage, Lyreleaf, a.k.a. Cancer Weed *Salvia lyrata* Eastern US	E–M Mint. Whorls of bilabial pale-lavender flowers with long corollas. 'Purple Knockout' and 'Purple Volcano' have burgundy foliage. Can spread aggressively.	Perennial Hardy to zone 6. Deer and drought resistant.	2 ft (60 cm) Sun to part shade, well-drained soil. Seed is easy to germinate. Long-blooming. Teas used to treat coughs. Allow space for air flow to prevent powdery mildew.	BB, HB, SB Long-tongued bees can reach the nectar, while smaller bees perform acrobatics to gather pollen. When bees enter the flower, stamens dust their hind ends with pollen.
Seablush *Plectritis congesta* Pacific coast from BC to CA	E Honeysuckle. Clusters of purple-pink blossoms, protruding stamens.	Perennial Hardy to zone 5.	2 ft (60 cm) Sun to part shade, well-drained soil. Seed germinates under cool conditions. Plant in fall. Scent may not appeal.	BB, HB, SB This Garry oak wildflower has special value to native bees. Important early source of nectar and pollen.
Shooting Star, Pretty, a.k.a. Prairie Shooting Star *Dodecatheon pulchellum* Western North America	E Primrose. Star-shaped flowers with protruding stamens. Pink or white flowers, depending on variety.	Perennial Hardy to zone 2.	1–2 ft (30–60 cm) Prefers partial shade; moist, well-drained soil rich in humus. Divide late winter or early spring. Seed needs cold stratification, cool soil and light to germinate. Water while blooming, then hold back to let plant go dormant.	BB Bees hang upside down to access nectar and pollen. An early bumblebee plant with unusual star-shaped flowers. Partner with blue camas and *Sisyrinchium* for a pollination powerhouse.
Solomon's Seal, Smooth *Polygonatum biflorum* East and central North America	E Asparagus. White flowers resemble small pairs of pantaloons. Berries are toxic. **Group with:** Lungwort, Virginia bluebells.	Perennial Hardy to zone 3.	1–2 ft (30–60 cm) Sun to full shade, sand, loam, medium to moist conditions. Divide and share. Suitable for pots, shade garden, below black walnut trees.	BB, SB Long-tongued bees such as bumblebees, digger bees, mining bees and large carpenter bees.
Stonecrop, Broadleaf *Sedum spathulifolium* Pacific coast from BC to CA	E Stonecrop. Small, shallow yellow flowers. **Group with:** Blue camas, nodding onion, salal.	Perennial Hardy to zone 3. Drought resistant.	8 in (20 cm) Full to partial sun, dry to moist, well-drained setting. Prefers poor, rocky soil. Easy to propagate from stem cuttings. Suitable for pots.	BB, HB, SB, BI Mining bees. Important larval plant for endangered butterflies. Essential roof-garden plant for pollinators.
Stonecrop, Woodland *Sedum ternatum* ON, eastern US	E Stonecrop. Small, shallow white flowers.	Perennial Hardy to zone 4.	4 in (10 cm) Sun to full shade in moist, well-drained understory. Divide or propagate cuttings.	BB, HB, SB, BI Mix with nodding onion and violets for a pollination triad. See broadleaf stonecrop.

PLANT NAME AND NATIVE REGIONS	PLANT FAMILY AND BLOOM	HARDINESS	MAXIMUM HEIGHT AND GARDEN NOTES	BENEFITS TO BEES*
Strawberry, Virginia *Fragaria virginiana* Canada, US	**E** Rose. Shallow white flowers. **Group with:** Borage, nodding onion.	**Perennial** Hardy to zone 3.	**4 in (10 cm)** Sun to shade, moist to medium soil. Seeds require cold, moist stratification. Spreads by runners.	**BB, HB, SB, BI** Cuckoo, green sweat, mason, mining, small carpenter and sweat bees.
Sundrops, Common *Oenothera fruticosa* ON, eastern US	**M** Primrose. Yellow cup-shaped flower. Can spread aggressively. **Group with:** Common milkweed, blazing star, Joe-Pye weed.	**Perennial** Hardy to zone 7. Drought resistant.	**1–2 ft (30–60 cm)** Sun, normal to sandy, medium to dry soil. Seed needs light and cold—sow fall or early spring. Divide every 2–3 years. Suitable for pots.	**BB, HB, SB, BI** Specialist pollinators: *Lasioglossum oenotherae* and the oil-collecting bee *Heriades*. Special value to native bees, butterflies, moths.
Sunflower, False, a.k.a. Heliopsis, Sweet Oxeye *Heliopsis helianthoides* Canada except north, east coast; central, eastern US	**M–L** Aster. Daisy-like flowers with yellow-green centres. Can spread aggressively. **Group with:** Anise hyssop, hoary vervain, perennial sage, purple prairie clover.	**Long-blooming Perennial** Hardy to zone 3.	**3–5 ft (90–150 cm)** Full to partial sun in clay-loam or rocky soils in moist to moderate conditions. Easy to start from seed. Divide in spring and fall. Deadhead for continued bloom. Control aphids with soapy water if needed. Suitable for pots.	**BB, HB, SB, BI** Bloom lasting 10–13 weeks supports cuckoo, green sweat, leafcutter, long-horned, mining, small carpenter and sweat bees. Monarch butterfly nectar plant.
Sunflower, Maximilian *Helianthus maximiliani* Eastern North America from MB to Texas	**M** Aster. Daisy-like flowers with yellow petals. Aggressive spreader. **Group with:** Canada goldenrod and fireweed.	**Perennial** Hardy to zone 4.	**40–80 in (1–2 m)** Sun, sandy to loamy soil. Seeds need cold, moist stratification. Suitable for wild meadow and hedgerow. Seeds are much loved by birds.	**BB, HB, SB, BI** Leafcutter, long-horned, mining and sweat bees. Honey plant. Late-season monarch butterfly plant. Specialist pollinators: Species of *Diadasia, Eucera, Melissodes* and *Svastra*.
Sunflower, Mexican *Tithonia rotundifolia* Southwestern US	**M–L** Aster. Dark-orange daisy-like flowers with light-orange centres. **Group with:** Other sunflowers, tall coneflowers.	Annual	**40–80 in (1–2 m)** Sun, average to fertile soil, dry to medium conditions. Start indoors in early spring or direct-sow in May. Seeds need some light to germinate. Plants may need support.	**BB, HB, SB, BI** Important source of late-season nectar and pollen. Leafcutter bees, hummingbirds and butterflies. Deadhead to encourage blooms.
Sunflower, Okanagan *Balsamorhiza sagittata* Western North America	**E** Aster. Yellow daisy-like flowers with orange centres. See woolly sunflower.	**Perennial** Hardy to zone 5. Drought resistant.	**1–2 ft (30–60 cm)** Sun, well-drained, sandy to loamy soils. Tricky to germinate from seed. Cold, moist stratification. Plants resent disturbance.	**BB, HB, SB, BI** Early-blooming open-access bee plant. Significant for native bee conservation, especially for restoration and reclamation sites.
Sunflower, Showy *Helianthus pauciflorus* AB to QC; central and eastern US	**M** Aster. Daisy-like flowers with yellow petals. Can spread aggressively. **Group with:** Canada goldenrod and fireweed.	**Perennial** Hardy to zone 5.	**40–80 in (1–2 m)** Full to partial sun in medium to dry well-drained soil. Cold, moist stratification. Suitable for wild meadow and hedgerow.	**BB, HB, SB, BI** Leafcutter and mining bees. Honey plant. Important butterfly plant. Birds love seeds. See Maximilian sunflower.
Sunflower, Woolly, a.k.a. Oregon Sunshine, Woolly Daisy *Eriophyllum lanatum* Pacific coast from BC to CA	**M** Aster. Daisy-like yellow flowers with yellow centres. Fragrant, silvery grey foliage. **Group with:** *Mahonia nervosa* and Okanagan sunflower.	**Perennial** Hardy to zone 5. Drought resistant.	**1–2 ft (30–60 cm)** Sun, preferably on a well-drained slope. Seeds need cold, moist stratification. Germination may take months. Best sown in the fall. Suitable for pots.	**BB, HB, SB, BI** Important open-access plant for a wide variety of bees and butterflies. Significant wildflower for native bee conservation.

PLANT NAME AND NATIVE REGIONS	PLANT FAMILY AND BLOOM	HARDINESS	MAXIMUM HEIGHT AND GARDEN NOTES	BENEFITS TO BEES*
Tickseed, Large-flowered *Coreopsis grandiflora* Eastern North America	M–L Aster. Daisy-like flowers with jagged-edged yellow petals, yellow centres. Aggressive spreader in moist conditions. **Group with:** Milkweed, hoary vervain.	**Perennial** Hardy to zone 4. Deer resistant.	2 ft (60 cm) Sun to dappled shade. Will grow in many soil types, but prefers well-drained, loamy soil. Seeds need cold, moist stratification. Self-seeding. Deadhead. Water regularly. Suitable for pots, rooftop.	BB, HB, SB, BI Cuckoo, leafcutter, long-horned, small carpenter, small resin bees. Specialist pollinator: coreopsis mining bee (*Melissodes coreopsis*).
Tickseed, Tall *Coreopsis tripteris* ON, QC, eastern US	M–L Aster. Yellow daisy-like flowers with yellow petals, brown centres. Aggressive spreader in moist conditions.	**Perennial** Hardy to zone 3. Deer resistant.	3–8 ft (0.9–2.4 m) See large-flowered tickseed. Tall tickseed blooms later than other tickseeds; plant more than one type.	BB, HB, SB, BI Essential open-access prairie pollinator meadow plant. See large-flowered tickseed.
Tidy Tips *Layia platyglossa* CA, southwestern US	M Aster. Yellow daisy-like flower with white tips on petals. **Group with:** Anise hyssop, poached egg flower.	**Self-seeding Annual** Drought resistant.	1 ft (30 cm) Sun to dappled shade, well-drained soil. In temperate zone, plant seed in fall; in cold zone, sow spring after last frost; germination needs light.	BB, HB, SB, BI Carpenter bees, mason bees and butterflies visit this flower in its native California. A pretty gap-filler to sow in any bee garden.
Turtlehead, White *Chelone glabra* Eastern North America	M–L Plantain. White flowers resemble turtle heads. *C. obliqua speciose* is pink. **Group with:** Blue vervain, swamp milkweed.	**Perennial** Hardy to zone 3. Deer resistant.	1–3 ft (30–90 cm) Full to partial sun in fertile soil, moist to wet conditions. Seed needs cold, moist, stratification. Plant in the fall. May need staking in shade. Suitable for pond edges.	BB, SB Bumblebees and long-horned bees climb right inside flowers to obtain nectar. Hummingbirds. Nectar contains a chemical (iridoid glycoside) that bees use to treat parasites.
Vervain, Blue *Verbena hastata* BC, SK, MB, ON, QC; much of continental US	M–L Verbena. Spikes of lavender-blue flowers. *V. urticifolia* is white. Long bloom time. Toxic. **Group with:** New England aster, swamp milkweed.	**Biennial or Short-lived Perennial** Hardy to zone 3. Deer resistant.	2–5 ft (60–150 cm) Sun to part shade in medium to soggy soils. Seeds need cold stratification, or plant in fall. Seeds need light to germinate. Self-seeding, but not aggressive. Suitable for pots, pond edge.	BB, HB, SB, BI Cuckoo, digger, green sweat, leaf-cutter, small carpenter and sweat bees. Long-tongued bees collect nectar and pollen. Specialist pollinator: verbena bee *Calliopsis verbenae*.
Vervain, Hoary *Verbena stricta* ON, QC, central US	M–L Verbena. Spikes of lavender-blue flowers. Long bloom time. **Group with:** False sunflower, goldenrod, milkweed, pearly everlasting, prairie clover, wild sunflower.	**Biennial or Short-lived Perennial** Hardy to zone 4. Deer and drought resistant.	2–4 ft (60–120 cm) Sun, well-drained sand to loam, medium to dry conditions. Seeds need cold stratification, or plant in fall. Seeds need light to germinate. Self-seeding, but not aggressive. Suitable for pots. Butterfly plant.	BB, HB, SB, BI Small carpenter, cuckoo, leaf-cutter, long-horned, miner, green sweat bees. Long-tongued bees collect nectar/pollen; short-tongued collect pollen. Specialist pollinator: Verbena bee (*Calliopsis verbenae*).
Violet, Early Blue *Viola adunca* BC, WA	E Violet. Light blue edible flowers with nectar guides and white centres.	**Perennial** Hardy to zone 4.	6 in (15 cm) Sun to shade, and medium to dry soil. Sow seeds in flats in early spring. Plant out in late summer. Great for pots.	BB, HB, SB Mason bees. Important early bumblebee flower. Specialist pollinator of violets: *Andrena violae*.
Violet, Labrador *Viola labradorica* Northeast North America	E Violet. Light-purple edible flowers with nectar guides and white centres.	**Perennial** Hardy to zone 5.	6 in (15 cm) See above. Note that that *V. riviniana* sold incorrectly as Labrador violet can be invasive.	BB, HB, SB Group violets with wild onions and columbine to please the bees. See early blue violet.

PLANT NAME AND NATIVE REGIONS	PLANT FAMILY AND BLOOM	HARDINESS	MAXIMUM HEIGHT AND GARDEN NOTES	BENEFITS TO BEES*
Yellow Rattle *Rhinanthus minor* Canada, AK, northern US	E–M Broomrapes. Yellow bilabial blossoms. Choose a subspecies for your bioregion. **Group with:** Blue-eyed Mary and seablush.	Self-seeding Annual	**20 in (50 cm)** Sun, well-drained, poor soil. Short-lived seeds require cold, moist stratification. Must be planted in fall. Used to suppress grasses in meadow and increase biodiversity.	BB, HB, SB Nectar accessed by long-tongued bees and nectar robbers. Top bumblebee conservation plant. Butterfly plant.
VINES				
Clematis, Virgin's Bower, a.k.a. Old Man's Beard, Devil's Darning Needles *Clematis virginiana* BC to NS, central and eastern US	E Buttercup. Flat white flowers with protruding stamens. Extremely toxic. **Group with:** Trees to climb, like black walnut, or sturdy shrubs such as ninebark.	Perennial Hardy to zone 3. Deer resistant.	**10 ft (3 m)** Partial shade, dry to normal conditions in rich, well-drained soil. Propagate from seed or bend vine and anchor into soil so it roots. Prune back hard in fall or early spring. Works for rooftop gardens. Needs support. Plant instead of *C. terniflora* or *C. paniculata*, which are invasive.	BB, HB, SB, BI Cuckoo, plasterer, masked, sweat bees. Native clematis can be honey plants where grown abundantly. Butterfly and hummingbird plant. Good vine to let climb a fence or trellis to add blossom density.
Honeysuckle, Orange Western Trumpet *Lonicera ciliosa* Pacific Northwest	E–M Honeysuckle. Long, red-orange tubular blossom. Berries are toxic. See twining honeysuckle.	Perennial Hardy to zone 5.	**10–30 ft (3–9 m)** Sun or partial shade, well-drained soil. Seeds must be cold stratified at least 3 months. Prune inferior shoots after blooming.	BB, SB See twining honeysuckle. Let it grow up ninebark, snowberry or oceanspray for a forage feast for bees.
Honeysuckle, Twining *Lonicera dioica* YT, NU, BC, Prairies, ON, QC, eastern US	E–M Honeysuckle. Red-orange tubular blossoms. Berries are toxic. **Group with:** Climbing roses; underplant with columbine.	Perennial Hardy to zone 3.	**10 ft (3 m)** Sun or partial shade, sandy soil. Propagate from cuttings started in pots. Needs trellis or support. Prune inferior shoots after blooming.	BB, SB Long-tongued bees and hummingbirds can reach copious nectar in deep blossoms.
Vetch *Vicia americana* YT, NWT, BC, Prairies, ON, QC, PEI, all but southeastern US	E–M Legume. Lipped blue, red and purple blossom with long nectar tube. **Group with:** Shrubs to climb.	Perennial Hardy to zone 3.	**1–3 ft (30–90 cm)** Sun or partial shade, moist soil high in clay. Inoculate seed. Seed should be a year old before planted. Self-seeding and spreads by rhizomes.	BB, HB, SB, BI Honey plant. Significant bumblebee plant. Nectar accessed by shorter-tongued bees when it is most abundant.

*Each pound per acre is equal to approximately 1.1 kilograms per hectare.

A cryptic bumblebee (*Bombus cryptarum*) rests on my hand in a garden in Cochrane, Alberta.

The New Buzz on Beekeeping

CHAPTER SEVEN

*C*oncerns brought on by Colony Collapse Disorder have led to an upsurge in hobby beekeeping across North America in both cities and rural settings. While well-meaning, this movement can sometimes lead to honeybees competing with wild bees and other honeybees for a paltry supply of pollen and nectar. Touted as a way to save the bees, beekeeping can, in fact, further contribute to their demise if there is a shortage of food to sustain them. Essential to keeping honeybees is ensuring there is enough forage to support both the honeybees *and* their native bee neighbours.

Most beekeepers care passionately about doing what's best for *all* bees. And by planting bee-supportive trees, shrubs, flowers and pasture, they can help improve the world for our ailing pollinators. Well-fed bees of all stripes are better equipped to fight diseases and pests so they are energized to mate, procreate and pollinate.

Opposite: The conscientious beekeeper will plant "offsets" for what their honeybees are consuming to ensure there is adequate pollen and nectar for neighbouring native bees.

Victory Beekeeping

World War II beekeepers were allowed a sugar ration to feed their bees. When the government found people were putting it in their tea instead of their hives, they dyed it green to make it less appetizing for the table. This resulted in unpalatable green honey—which brought this practice to an abrupt end!

The Beekeeper's Garden: Keep the Nectar Flowing, Honey!

The exemplary beekeeper feeds her bees with pesticide-free pasture and orchards that bloom from early spring to fall. She gives them fresh water and grows herbs so they can count on the essential oils to repel bacteria and hive pests. Honeybees need trees like balsam poplar (*Populus balsamifera*) or pine (*Pinus* spp.) to provide the resins they require to make the propolis that seals and protects the hives. Other key plants include pussy willow to help bees build up spring brood and goldenrod to support them in stocking up the hive for winter. Bee pasture rich in nectar and pollen can be sown in succession to keep the honey flowing. The responsible keeper also provides extra food and habitat for the surrounding native bees busily working to provide for their families, ensuring a feast of forage for everyone. See opposite page for a sample beekeeper's garden plan.

Brian Campbell is a sweet and humorous beekeeping and gardening savant. He teaches beekeepers to put the needs of the bees above profit.

committed to the conservation of native bees—has generously shared his skills and passion with me over many years. He has taught a generation of beekeepers to put the health of the bees first, and that the key to keeping all bees healthy is providing top-quality forage . . . and enough of it for *all* the inhabitants in the bee "neighbourhood."

To connect with a mentor, volunteer to help an established beekeeper in your area—they can always use an extra pair of hands in the apiary, and this is the

WHY EVERY BEEKEEPER DESERVES A MENTOR

Any beginning beekeeper will benefit immensely from a good mentor. Brian Campbell—originator of the Blessed Bee Community Apiary and Bee School in Vancouver, BC, and an organic master beekeeper

Hives for Humanity

Sarah Common, a social worker in Vancouver's Downtown Eastside and co-founder of Hives for Humanity, believes that beekeeping can be highly therapeutic if done slowly, with intent and with a spirit of community and collaboration.

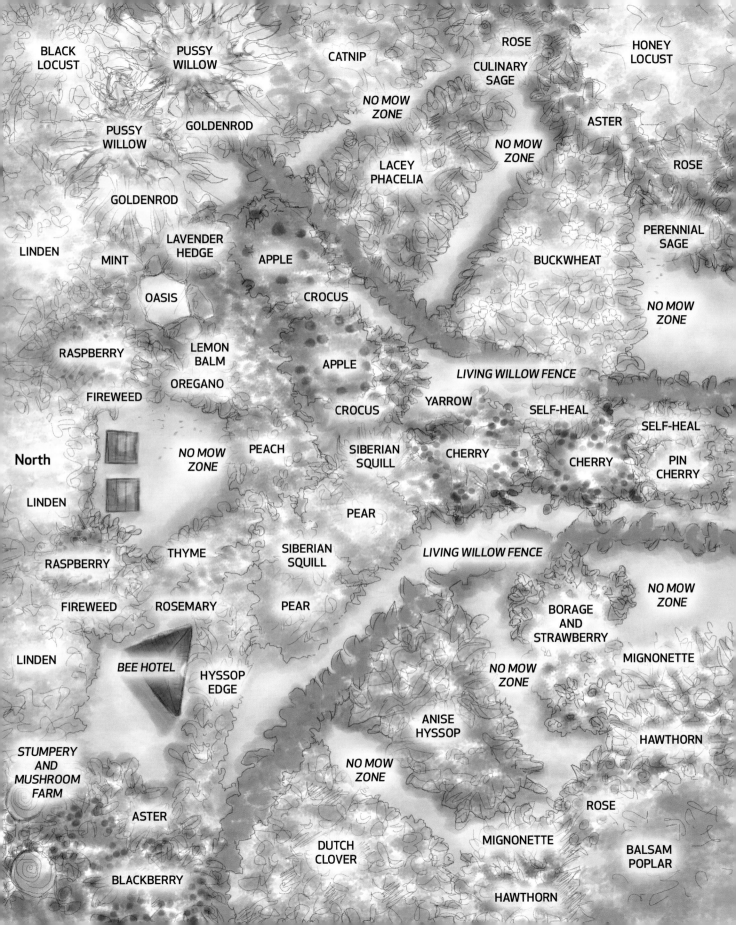

At the Root of Beekeeping

Early beekeeping in northern Europe was very forest-based, with the first beekeepers not really keepers so much as climbers who harvested wild honey found in hollowed-out trees. Trees inhabited by honeybees were called *honigbaum* and tree-stump hives *bienenstock*, the word for beehives used today in Germany.

Above: This bee (centre left) has collected propolis from a tree with her mandibles and transferred it to her back legs. It helps keep the hive hygienic. **Opposite, clockwise from top left:** Crimson clover is a beautiful addition to fall bouquets. Yellow-faced bumblebees forage in vetch climbing up a snowberry bush. Judging by the number of stacked hive boxes, my grandpa Fred was an extremely successful beekeeper in the aptly named municipality of Bounty, Saskatchewan.

best way to get started. Gift them with some dried sage and lavender from your garden to use in their smoker. Mead is also another traditional offering to give to your favourite "beek" (slang for beekeeper). In an urban setting, there are also growing opportunities to help out with community hives without increasing the honeybee footprint in an area with limited nectar resources.

GIVING BACK TO BEES

Bees can provide a connection to the natural world, but no hive is an island; consider that your honeybees can impact native bees within a radius of roughly 4 miles (6 km). When my grandpa Clark was a beekeeper in the decade after World War II in Bounty, S K, he had no problem filling supers with pounds of honey. That was because he planted shelterbelts with shrubs that provided nectar to supplement the flowering crops (such as alfalfa) that provided bee pasture, along with the many wildflowers that grew then in the nearby prairie meadows. He practised sustainable beekeeping with one or two hives, recognizing just how much forage is needed for each colony. Even before World War II, pioneering B C botanist John Davidson painted a picture of

just how much forage is needed for honeybees—37,000 loads of nectar for just 1 pound (455 g) of honey!

And it's not enough to just consider the honey when planting for bees. It's important to factor in the blossoms needed to create the wax comb—the infrastructure of the hive—along with the calories to sustain all that foraging, flying, royal-jelly making and bee dancing. Flowers also provide the pounds and pounds of pollen to feed the brood. For generations, honeybees have given humanity the gifts of sweetness and light, but we must provide them with a sufficient bounty of flowers in our gardens, hedgerows, fields and pastures. Think of this as good bee karma, deeds imperative to the future of our planet.

J.F.C.   Aug. 1930.

I have fun showing students the trick I learned from beekeeper Brian Campbell: pressing gently on the backs of drones to hear them buzz.

THE GOLDEN RULES FOR SUSTAINABLE BEEKEEPING:[35]

- Be aware of how many honeybee hives are in a radius of roughly 4 miles (6 km), so you do not overload the area's resources.
- Learn basic bee biology and keep up with scientific advancements in beekeeping.
- Use organic methods to prevent and treat honeybee diseases; otherwise they may spread to native bees or other honeybee colonies.
- Only take sustainable amounts of hive products, such as honey, wax and propolis.
- Refrain from loading up areas of wilderness with honeybees.
- Plant forage to offset the nectar and pollen your honeybees take from the environment.

An average hive can consume 300 to 500 pounds (136 to 227 kg) of nectar and 100 pounds (45 kg) of pollen per year.

SUSTAINABLE HIVE HARVESTING

Organic beekeepers and biodynamic apiarists make sure they leave enough honey in the hive so the bees do not have to rely on sugar water, which lacks the vitamins, minerals and antibacterial action provided by honey.[36] In addition to optimizing the health of the honeybees, leaving honey in their hive means they need less nectar from the field, which leaves more forage for native bees. The same goes for other hive products, such as pollen, royal jelly and propolis, which are best left in the hive to support the bees, since they have more health benefits for bees than for humans.

Pollen Power

While pollen is sold as a nutritional supplement for humans, sustainable beekeepers leave it in the hives as a superfood for baby bees. Any overharvesting of pollen compromises the health of the hive and requires the honeybees to harvest more to make up for what is missing—which, again, means less pollen for native bees.

Royal Jelly

Royal jelly is secreted by nurse bees and fed to all larvae for three days. It was previously thought to be the main food of aspiring larvae that would be queens, but new research brings the whole purpose of regal bee slime into question.[37] Whatever its use, natural

Raw, Runny and Rocky

Raw honey has been used as a treatment for topical wounds from ancient Egypt to the battlefields of World War II, and is still used in veterinary medicine. "Raw" honey has not been heated above hive temperature, nor has it been filtered to remove pollen. Pasteurizing honey by heating it above hive temperature destroys its medicinal qualities, while fine filtering strips micronutrients contained in the pollen grains.

Sometimes your nourishing raw honey can undergo a chemical change called "crystallization," where it may "seize up" and develop a gritty texture. Left like this for too long, it will start to ferment. This natural process can be easily reversed by gently heating the honey. Using a candy thermometer, warm a small pot of water to hive temperature (95–115F/35–46C). Then turn off the heat and set the jar in the water bath until it becomes runny again. Honey is best stored in glass for this purpose.

The burr comb at the end of this frame contains drone cells built for developing male bees.

beekeepers leave it in the hive for the health of the bees.

Potent Propolis

Propolis is the most powerful medicine in the beehive. Honeybees have learned to gather saps, latex and other plant materials to make a bee glue to seal the hive and make the surfaces of their home antibacterial and antifungal. Bees also mix propolis with wax to repair comb and modify their homes. New evidence found by scientist Marla Spivak shows that it is one of the keys to keeping hives healthy; she advises beekeepers to rough up the inside of the hive to encourage its inhabitants to coat it with propolis.[38] Plant resin-bearing trees, such as balsam poplar and pine, to give back to the bees. You can also make your own propolis-like salve, called "Balm of Gilead," from poplar bud resin and leave the bee glue in the hive for the health of the colony.

HONOURING HIVE WAX

When master beekeeper Brian Campbell talks about honeybees, he reminds us that there are four castes in the hive: queens, worker bees, drones and the wax comb. Wax is an integral part of the superorganism of the hive, where *everything* happens on the comb. Without wax, there would be no colony, says Campbell. The comb functions as the skeletal structure of the hive, an integral part of the nervous, digestive and immune systems of the superorganism, as well as its communication network and uterus. It can serve a queen throughout her life of up to six years as well as continuous generations in the same location. The comb structure of the hive takes a tremendous investment of time and energy—with worker bees painstakingly extruding it through plates on their bellies from wax-producing glands in their abdomens—and should be honoured accordingly.

A single pound (455 g) of wax also requires almost 8 pounds (3.6 kg) of honey, which takes over 20 pounds (9 kg) of nectar, so the more you leave in the hive, the more energy the bees can devote to foraging for food, and the less pressure it puts on the resources in your area. Some of the old comb does need to be cycled out of the hive, however, because honeybees are sensitive to contaminates (neonics, fungicides and other toxins) that accumulate in the wax and pollen. New wax is a translucent ivory colour, while old comb is almost black and mostly made of bee-cocoon remnants and propolis, with hardly any wax left. To support hive health, Campbell cycles a third of the wax out every year to reduce the buildup of environmental contaminants embedded in the comb. The crush method of honey extraction (crushing the wax and allowing it to flow out of the comb) accomplishes a similar portion of wax renewal. Or some beekeepers choose to sell honey in the comb to preserve its flavour in the individual cells—this also cycles out wax. Used wax can be made into candles if it does not contain pesticides. Otherwise, it needs to be safely discarded.

Wartime Beekeeping

Beekeeping in World Wars I and II saw women entering the field as a way to supplement their families' wartime diet and/or income. In her book *Beeconomy*, Tammy Horn relates a cautionary tale of the pitfalls brought on by the gold-rush mentality of wartime beekeeping. In 1917, a Wisconsin woman named Mathilda Candler began beekeeping with two hives.[39] She flourished as a beekeeper, building her apiary to 49 colonies, and hit the jackpot when her honey sold for a total of $600 one year. Suddenly, everyone in the neighbourhood thought it would be a good idea to keep bees, creating such a density of hives that there wasn't enough forage to go around . . . and many colonies failed, with the newcomers with gold-rush fever returning home in debt for the cost of their hives. After World War I, women continued keeping bees to meet the demand for the products of the hive, but by World War II, the US government had put a ceiling on wax and honey prices, so beekeeping was no longer considered the gold mine it once was.

Opposite, clockwise from top left: In the United States during World War II shortages, old-timers taught folks how to make hives from straw. Nectar-rich clover interplanted with kale fixes nitrogen in the soil, suppresses weeds and prevents erosion. Sweat bees (*Lasioglossum*), masked bees (*Hylaeus*) and others love to forage on carrot blossoms. Phacelia interplanted with crimson clover helps improve soil and prevent periods of nectar and pollen dearth.

The cosmetics industry was dependent on maintaining a supply of beeswax in World War II, but beeswax was also needed for coating planes and other utilitarian uses. As a clever tactic, lipstick was marketed as "patriotic" and took on the characteristic bullet shape we know today. It was one luxury product exempted from being rationed.[40]

BEE PASTURES WITH BENEFITS

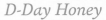

D-Day Honey

Normandy-based beekeeper Ed Robinson sells D-Day Honey. The beekeeper/historical-tour guide has a 5-acre (2-ha) bee pasture he's seeded with wildflowers with some special assistants. He dumps roughly 18 pounds (8 kg) of seeds into three pails for his neighbour's children to go forth and distribute. "It's a match made in heaven," he says. "A three-year-old with a bucket of wildflower seeds."

The word "pasture" is an agricultural term meaning land covered with forage for grazing animals. The term "bee pasture" traditionally refers to honeybees, but needs to be opened up to include plants for native

Below: Hives for Humanity honey is labelled according to the neighbourhood where it was harvested. Honey gets its unique flavour from the flowers it is made from. **Opposite:** A yellow-faced bumblebee forages on Dutch clover. If you look very closely, you can see a mite hitching a ride. JACK TUPPER PHOTO

bees. Some bee pastures are composed of wildflowers and grasses, while others consist of cultivated cash or cover crops. There are three key ways you can create pasture for bees: dig up a patch of lawn to seed clover, plant meadows in field margins, and grow biodiverse cover crops rich in nectar and pollen.

Bee-wise Farm Models

Integrating bee pasture into food and livestock forage crops is an important large-scale strategy for saving bees. Revolutionizing the monoculture model of food production this way is crucial as we try to squeeze more nutrition out of less land for a growing global population. A key strategy to make agriculture more sustainable is to devote at least one-third of the land mass on a farm to margins of native and near-native plants for native bee forage. Native perennials (and select self-seeding annuals) grown on undisturbed land will support the guilds of native bees, allowing farmers to become less dependent on honeybee rentals and benefit from increased yield and profits.

Canada is the world's top grower of canola, which is also the main honey plant for many western provinces. While canola is a good nectar plant, the seeds are often genetically modified and coated in neonicotinoids—*not* good for bees. Camelina (*Camelina sativa*), on the other hand, is an oilseed crop loved by honeybees and proving to be more sustainable, since it requires less fertilizer and fewer pesticides and energy inputs than canola. It also attracts syrphid flies and native sweat and masked bees. Both camelina and canola fields benefit from bee-pasture margins equalling at least a third of the land mass[41] to support the native bees that pollinate these crops.[42]

GIVING BACK TO THE SOIL WHILE PAYING IT FORWARD FOR THE BEES

If you are considering turning your front lawn into a food garden, amend the soil before you start planting your Victory veggies. You'll multiply the benefits if you prepare the earth by planting a bee pasture. The plants will loosen the soil, repel nematodes, add nutrients—and feed bees! Try this mix for an optimal synergy of bee-powered bliss:

- 30 percent phacelia
- 20 percent buckwheat
- 20 percent crimson clover
- 10 percent borage
- 5 percent calendula
- 5 percent coriander
- 5 percent white mustard
- 3 percent daikon radish
- 2 percent dill

Consider dividing your garden in half and planting this mix on alternate sides year to year to give back to the soil while giving back to the bees.

A bumblebee carries three layers of pollen from different plants, including blue from this lacey phacelia.

The Pin and Thrum of Buckwheat Flowers

Buckwheat plants have two distinct kinds of flowers—pin and thrum. Pin flowers have shorter male parts (stamens) and the longer female part (style). Thrum flowers are the opposite, with longer stamens and a shorter style. Honeybees tend to go for thrum flowers because they produce the nectar, so native bees, such as bumblebees, which visit both the male and female flowers, are very important pollinators of this crop and essential for successful seed set.

QUANTITY MEANS QUALITY

In 1920, Frank C. Pellett published *American Honey Plants*, which gives solid advice to beekeepers to this day: "The ideal situation for beekeeping is one where there are at least three plants which yield surplus honey in considerable quantity, and which bloom at different periods. Beside the main sources, there should be a great variety of minor plants yielding both pollen and honey throughout the season to support the bees between the main flows. In such a situation, there is seldom an entire failure of the honey crop; and, in good years, the beekeeper fares well, indeed."[45]

TRIPLE SWEET TREATS: HONEYBEE PASTURE SUPERFOODS

A top honey plant, buckwheat (*Fagopyrum esculentum*) comes with a grocery list of advantages for gardens and fields, including attracting scores of beneficial insects and making a fast-growing green manure. It also provides highly nutritious seeds that can be ground into gluten-free flour with a food processor. Lacey phacelia

Scientist Bernd Heinrich estimates that in ideal conditions, without forage competition, a single bumblebee can extract enough energy from one fireweed flower to give her enough energy to forage for fourteen minutes. Two or three flowers give a honeybee a whopping full load of nectar.[48]

Fireweed supported feral honeybees when they were first introduced to British Columbia in the 1800s.

This black-tailed bumblebee (*Bombus melanopygus*) has corbiculae packed full of the brown pollen from red clover.

fodder that feeds wildlife, livestock and bees. Its edible pink pea-like flowers are marked with vivid nectar-guide stripes and reputed to attract 10 times more bees than white Dutch clover.[46] As the flowers begin blooming in May, sainfoin is a good orchard plant, extending the late-spring honey flow. Just be prepared for honeybees to favour it above just about anything else blooming at the same time.

THE LUCKY CLOVER CLUB

Poet Emily Dickinson wrote that "To make a prairie it takes a clover and one bee . . . And revery."[47] Bees cannot live on reverie alone, but clover will help them stock their pantries. When planting clover in your bee pasture, choose varieties best suited to your site. Three Old World clover classics are white Dutch (*Trifolium repens*), red (*T. pratense*) and crimson (*T. incarnatum*). All fix nitrogen in the soil and, when planted together, provide long-blooming nectar-filled florets for short- to

(*Phacelia tanacetifolia*) is also one of the best honey plants around, providing striking blue-purple pollen and pumping out nectar throughout the day. Since it belongs to the borage family, it can easily fit into crop-rotation plans. Sainfoin (*Onobrychis viciifolia*) comes from the Latin *sanum foenum,* meaning "healthy hay." This Old World gem is worth reviving for a perennial

This wax comb is pale because it is freshly made. The black cell in the centre is packed with medicinal poppy pollen.

long-tongued bees. White Dutch has the shortest corollas, and is a good match for honeybees and other bees of medium or small size. Red has the longest corollas of the trio, making it tops for bumblebees. It's also the best pick for poor soils, as the long taproots pull up minerals and loosen the earth. And for improving soil, preventing erosion and suppressing weeds in your garden beds, crimson doesn't become weedy like the other kinds of clover, which can choke out less assertive plants. Meanwhile, it is superior for honeybees, too, and feeds digger, large leafcutter and mason bees.

A TRIO OF BOTANY JOHN'S WILDFLOWERS

As far back as the roaring twenties, pioneering B C botanist John Davidson recommended local native plants suitable for honeybees, cautioning beekeepers that "you can have no more bees than the flora will support; apart from insect parasites, the flora is the limiting factor." [49] He noted that fireweed (*Chamerion angustifolium*) is a key reason honeybees flourished when they were brought to North America. As trees were cut to make room for development, acres of purple fireweed flowered, providing ample forage for bees and honey

for early settlers. Another plant favoured by "Botany John," spotted bee balm or horsemint (*Monarda punctata*), historically a bee plant native to hot, dry regions, is ideal for xeriscape (water-conserving) gardens. He also advocated later-blooming native asters as fall forage for bees building winter stores, claiming "much of the value credited to goldenrod belongs to asters." [50]

BEE PASTURE *plants*

A SEASONAL PLANTING CHART

*F*lowers can provide a healthy meal plan for bees while boosting soil health. Develop blends that complement each other, or interplant rows of single varieties to break up the monoculture model. Choose disease-resistant cultivars and plants native or nearly native to your region. Create a mix accessible to both long- and short-tongued bees to ensure that your garden provides rewards to a wide spectrum of pollinators.

PLANT ORIGIN

OW: Old World (indigenous to Europe, Asia or Africa)

NW: New World (indigenous to the Americas and their islands)

BLOOM PERIOD

E: Early

M: Mid-season

L: Late

SUC: May be seeded in succession

HARDINESS refers to the coldest zone the plant is hardy to.

DEER AND DROUGHT RESISTANCE are most reliable once the plant is established.

BEES ATTRACTED

BB: Bumblebees

HB: Honeybees

SB: Solitary bees

BI: Beneficial insects

PLANT	BLOOM AND PLANT FAMILY	HARDINESS	MAXIMUM HEIGHT AND BEE PASTURE NOTES	BENEFITS TO BEES*
Alfalfa, a.k.a. Lucerne *Medicago sativa* **OW**	**M–SUC** Legume. Loose clusters of purple or yellow flowers open from base to top.	**Annual or Perennial** Hardy to zone 5.	**30 in (76 cm)** Prefers sun and light soil. Nitrogen-fixing cover crop with long roots and a week-long bloom. Nutritious fodder for livestock.	**BB, HB, SB, BI** Top honey plant: 180–445 lbs/ac. Trip-pollinated by alfalfa leaf-cutter, bumblebee, mining, sweat bees. Honeybees rob nectar.
Aster, Smooth *Symphyotrichum laeve* **NW:** YT, BC, Prairies, ON, continental US except CA	**M–L** Aster. Light-purple daisy-like flower with yellow centre. **Group with:** Goldenrod.	**Perennial** Hardy to zone 4. Deer and drought resistant.	**2–4 ft (60–120 cm)** Sun to part shade in mix of sand, loam, clay. Medium to moist conditions. Long-blooming.	**BB, HB, SB, BI** Honey potential (*Symphotrichum* spp.): 30–50 lbs/ac. Cuckoo, leafcutter, long-horned, mining, green sweat, small carpenter bees. Important fall source of pollen and nectar.
Bee Balm, Spotted *Monarda punctata* **NW:** ON, QC, eastern US	**M–L** Mint. Tiers of small exotic flowers with showy bracts. **Group with:** Aster, mountain mint.	**Short-lived Perennial** Hardy to zone 3. Deer and drought resistant.	**1–3 ft (30–90 cm)** Barely cover seed; needs light to germinate. Sun, sandy soil, dry to moderately moist conditions.	**BB, HB, SB, BI** Prolific honey plant. Most accessible *Monarda* to a variety of bees. Favourite of long-horned, mining, plasterer, sweat bees.
Boneset, Common, *Eupatorium perfoliatum* **NW:** MB, ON, QC NS, NB, PEI, eastern US	**M–L** Aster. Loose clusters of fringed white flowers. **Group with:** Blue vervain.	**Perennial** Hardy to zone 3. Drought resistant.	**2–5 ft (60–150 cm)** Full to partial sun, medium to wet soil. Cold stratify seed or sow in fall. Seeds need light to germinate.	**BB, HB, SB, BI** Leafcutter, mining, small carpenter, sweat bees. One of the best *Eupatorium* spp. for honeybees.
Borage *Borago officinalis* **OW**	**E–M** Borage. Blue star-shaped blossoms, black stamens. For white flowers, *B. officinalis* 'Alba'. *Seeds contain toxins. Avoid ingesting this herb if pregnant or suffering from a liver ailment.* **Group with:** Crimson clover, lacey phacelia.	**Self-seeding Annual** Deer resistant.	**2 ft (60 cm)** Prefers sun to part shade. Thrives in poor soil if good drainage. Direct-sow. Long taproot. Water well to keep up nectar supply. Green manure.	**BB, HB** Top honey plant: 200 lbs/ac and 60–160 lbs/ac blue-grey pollen. Refills with nectar every 2–5 minutes, unlike most plants. Significant bumblebee plant. Buzz-pollinated.
Broad Beans, a.k.a. Fava Beans *Vicia faba* **OW**	**E–M–L–SUC** Legume. White flowers with purple nectar guides and black-purple landing spots. Nutritious edible beans.	**Annual**	**40 in (1 m)** Sun and moist, well-drained soil. Sow when soil is cool. Nitrogen fixer. Choose open-pollinated varieties. Susceptible to aphids, slugs. May need to be staked.	**BB, HB, SB** Nectar accessible to long-tongued bees and nectar robbers. Extra floral nectaries. Bees help with seed set. Good fall forage for bees when little else blooms.
Buckwheat *Fagopyrum esculentum* **OW**	**M** Buckwheat. Clusters of small white flowers, tinged with blush pink. Dark-brown seeds highly nutritious. *Not suitable for livestock. Avoid eating sprouts, which are toxic. Avoid tartary buckwheat (F. tataricum), a noxious weed.* **Group with:** Alfalfa, crimson clover, sunflower.	**Frost-tender Self-seeding Annual**	**2–5 ft (60–150 cm)** Yields nectar best on light, sandy soils. Tolerates acidity. Broadcast seed onto lightly raked soil in a sunny site, covering lightly. Fast-growing cover crop. Dig under 10–12 weeks after blooming. Smothers weeds.	**BB, HB, SB, BI** Important honey plant: 45–445 lbs/ac. Produces nectar from morning to midday. Dark, pungent honey. Cool, moist conditions optimize nectar production. Japanese buckwheat particularly nectar-rich.

*Each pound per acre is equal to approximately 1.1 kilograms per hectare.

Milkweed is a trifecta pollinator plant, providing nectar for bees, hummingbirds and migrating monarch butterflies. Try to choose a species that is native to your bioregion.

PLANT	BLOOM AND PLANT FAMILY	HARDINESS	MAXIMUM HEIGHT AND BEE PASTURE NOTES	BENEFITS TO BEES*
Catnip *Nepeta cataria* OW	M–L Mint. Spears of bilabial white flowers with purple nectar guides and protruding stamens. Can spread aggressively. **Group with:** Basil, purple perilla, mint.	**Short-lived Perennial** Hardy to zone 3. Deer and drought resistant.	3 ft (90 cm) Sun, a well-drained, limey soil with high pH. Sow spring or fall; seed needs light to germinate. Good bee pasture for vineyards. Helps improve soil around the vines. Self-seeding.	BB, HB, SB, BI Honey potential: 90–445 lbs/ac. Cuckoo, digger, leafcutter, sweat bees.
Clover, Crimson *Trifolium incarnatum* OW	E–M–L–SUC Legume. Spikes of blood-red bilabial flowers. Can displace native vegetation. **Group with:** Daikon radish, hairy vetch, lacey phacelia.	**Self-seeding Annual**	8–20 in (20–50 cm) Sun, well-drained soil. Fixes nitrogen. Green manure and livestock fodder. Suitable for one cutting of hay.	BB, HB, SB, BI Honey potential: to 22 lbs/ac. Sugar concentration 31–60 percent. Superior for honeybees. Digger, large leafcutter, mason bees.
Clover, Dutch, a.k.a. White Clover *Trifolium repens* OW	M–L–SUC Legume. White blossoms, some pinkish. Young leaves, flowers edible if steamed. Tea from flowers used to treat colds. Can be invasive in small gardens.	**Perennial** Hardy to zone 4. Drought resistant.	8 in (20 cm) Sun and well-drained soil. Overseed lawn in spring. Nitrogen fixer; green manure. Key lawn substitute to support bees. Self-seeds and spreads by stolons.	BB, HB, SB, BI Top honey plant: 45–90 lbs/ac. Sugar concentration to 64 percent. Brown high-protein pollen. Cuckoo, leafcutter, sweat, mason, mining, resin bees.
Clover, Prairie *Dalea purpurea* **NW:** SK, MB, ON, central US	M Legume. Spikes of purple-pink or white flowers, protruding stamens and bright yellow-orange pollen. **Group with:** Anise hyssop, lacey phacelia.	**Perennial** Hardy to zone 3. Drought resistant, especially white-flowered *D. candida*.	1–3 ft (30–90 cm) Sun, moist or dry conditions. Easy to grow from seed in warm soil. Deep taproot. Fixes nitrogen. Treat with inoculum. Suitable for green roof, prairie meadow, roadside. May self-seed.	BB, HB, SB, BI Essential bee wildflower. Significant bumblebee plant. Cuckoo, green sweat, leafcutter and other sweat bees.
Clover, Red *Trifolium pratense* OW	E–M–L–SUC Legume. Flower heads made of florets with long pink-mauve to white nectar tubes. **Group with:** White lupin for bumblebees.	**Perennial** Hardy to zone 3.	8–30 in (20–76 cm) Part shade to sun in well-drained fertile soil with pH 6–7. Fodder for livestock. Green manure; 3 successive crops a year can be sowed. Fixes nitrogen. Self-seeding.	BB, HB, SB, BI Honey plant; honey potential 45–90 lbs/ac. Dark-green to reddish-brown pollen. Bumblebees, digger, large leafcutter, mason, mining, sweat bees. Short-tongued bees access nectar when corollas filled to brim.
Cup Plant *Silphium perfoliatum* **NW:** ON, QC, eastern US	L Aster. Goldfinches eat seeds and drink water from yellow cups. Can be invasive.	**Perennial** Hardy to zone 6.	3–8 ft (0.9–2.4 m) Full or partial sun in loamy soil, moist conditions. Seeds need cold, moist stratification unless planting in fall. Self-seeds. Forage for livestock.	BB, HB, SB, BI Cuckoo, digger, green sweat, leafcutter, long-horned bees, small carpenter bees.
Daikon Radish *Raphanus sativus* OW	M Brassica. White, 4-petalled flowers. **Group with:** Oats, turnips.	**Self-seeding Annual**	1–2 ft (30–60 cm) Sun. Improves soil tilth, suppresses weeds and nematodes.	BB, HB, SB, BI When grown for seed, radishes are good nectar plants. Honey is spicy.
Dock, Prairie *Silphium terebinthinaceum* **NW:** ON, eastern US	L Aster. Yellow daisy-like flowers. **Group with:** Coreopsis, milkweed.	**Perennial** Hardy to zone 4.	3–10 ft (0.9–3m) Similar to cup plant, but prefers drier conditions. Self-seeds.	BB, HB, SB, BI See cup plant. Special value for native bees. Provides nesting material for native bees. Hummingbird plant.

PLANT	BLOOM AND PLANT FAMILY	HARDINESS	MAXIMUM HEIGHT AND BEE PASTURE NOTES	BENEFITS TO BEES*
Fireweed *Chamerion angustifolium* **NW:** Canada, US except southeast	**M** Evening primrose. Spears of purple flowers with protruding stamens. White cultivar is 'Album'. Can spread aggressively. **Group with:** Cup plant, goldenrod, Joe-Pye weed.	**Perennial** Hardy to zone 2.	**2–6 ft (60–180 cm)** Sun, dry to moist soil. Easily started from seed. Divide rhizomes early spring or fall. Self-seeding. Suitable for large pots, hedgerow, meadow. Antibacterial, anti-inflammatory herbal. Good for habitat restoration.	**BB, HB, SB, BI** Honey potential: 180–715 lbs/ac or more. One of the most important Canadian heritage honey plants. Leafcutter and other long-tongued bees access nectar. Smaller-tongued bees harvest blue or teal pollen.
Goldenrod, Wrinkleleaf *Solidago rugosa* **NW:** Eastern North America	**L** Aster family: Spikes of small yellow flowers for late-season pollen. Aggressive spreader.	**Perennial** Hardy to zone 4.	**3–4 ft (90–120 cm)** Sun, medium to wet soil. Cold stratify seeds or plant in fall, barely covering. Self-seeding.	**BB, HB, SB, BI** Honey potential (*Solidago* spp.): 25–50 lbs/ac. Mining, large carpenter, sweat bees and more.
Hyssop, Purple Giant *Agastache scrophulariifolia* **NW:** ON, eastern US	**M–L** Mint. Spikes of pale-purple bilabial flowers that produce nectar all day, unlike most plants.	**Perennial** Hardy to zone 5.	**80 in (2 m)** Sun to part shade, moderately wet to medium dry soil. Seeds need cold stratification, light to germinate. Self-seeding.	**BB, HB, SB, BI** Top honey plant. Leafcutter bees. Special value to native bees, especially bumblebees.
Lavender *Lavandula angustifolia* **OW**	**M–L** Mint. Spikes of purple flowers with short nectar tubes. Cultivar flowers pink or white.	**Perennial** Hardy to zone 6. Deer and drought resistant.	**20 in (50 cm)** Growing lavender on a large scale involves propagating from cuttings. (Plan for a 50 percent success rate.) Sun, well-drained soil.	**BB, HB, SB, BI** Honey potential: 45–90 lbs/ac. Sugar concentration 33 percent. Significant bumblebee plant.
Lupin, White *Lupinus albus* **OW**	**E–M** Legume. Spikes of white pea-like blossoms. Most lupins toxic to humans, livestock.	**Annual**	**20 in (50 cm)** Sun, moderate to poor, well-drained soil. Nitrogen fixer. Deep taproot.	**BB, SB** Plant as nectar offsets for bumblebees. Butterfly plant.
Maralroot *Leuzea carthamoides* 'Lujza' **OW**	**E–M** Aster. Large purple-fringed blossoms similar to thistles. Blooms second year. Roots harvested after 2 years for medicinal use. *Contraindication for high blood pressure.*	**Perennial** Hardy to zone 2.	**30–60 in (76–152 cm)** Sun and deep, well-drained soil. Sow early spring. Cold stratification. Heat inhibits germination. Propagate by root splitting April or August. Taproot. Needs air flow; give it space. Forage for livestock.	**BB, HB, SB, BI** Significant honey plant: 27–100 lbs/ac. Important bumblebee herb. Originating in Siberia, can be grown in a short garden season.
Milkweed, Butterfly *Asclepias tuberosa* **NW:** ON, QC, CA, central and eastern US	**M** Dogbane. Clusters of bright-orange flowers. The pollen is made up of chains, rather than grains. Toxic.	**Long-blooming Perennial** Hardy to zone 4. Deer and drought resistant.	**2–3 ft (60–90 cm)** Sun to part shade in sandy to loamy soil, dry to medium conditions. Easy to start from seed sown directly into soil. Deep taproot, does not like to be moved. Divide roots in early spring or fall.	**BB, HB, SB, BI** Honey potential (*Asclepias* spp.): 120–250 lbs/ac, depending on soil quality. Blooms 8–10 weeks for cuckoo, leafcutter, mining, small carpenter, small resin, sweat bees.
Mustard, White *Sinapis alba* **OW**	**E–M** Brassica. Clusters of small yellow flowers with 4 petals. Rotate with crops other than brassicas.	**Annual**	**1–2 ft (30–60 cm)** Dig under 3 weeks before replanting pasture. Some nitrogen. Green manure. Suppresses weeds. Edible seeds.	**BB, HB, SB, BI** Honey potential: up to 90 lbs/ac. Sugar concentration up to 60 percent. Good-quality yellow pollen.
Pea, Austrian Winter, a.k.a. Hairy Vetchling *Lathyrus hirsutus* **OW**	**E** Legume. Purple or blue pea-like flower. Pink-flowered *L. sylvestris* 'Wagneri' a heritage honeybee forage.	**Annual** Drought resistant.	**30 in (76 cm)** Ground cover. Fixes nitrogen. Food crop in Asia and Africa. (Names can be confusing: *Pisum arvense* also known as Austrian winter pea, another bee-friendly cover crop.)	**BB, HB, SB** Significant nectar source for long-tongued bees. Large carpenter bees. Butterfly plant. Train vines up a trellis to increase blossom density for bees.

PLANT	BLOOM AND PLANT FAMILY	HARDINESS	MAXIMUM HEIGHT AND BEE PASTURE NOTES	BENEFITS TO BEES*
Pea, Grass *Lathyrus sativus* **OW**	E Legume. Small purple, pink, red, white or blue pea-like flowers. Contains neurotoxins that are being bred out.	Annual Drought resistant.	**2 ft (60 cm)** Tolerates extreme growing conditions including drought and flood. Fixes nitrogen. Food crop in Asia and Africa.	BB, HB, SB See Austrian winter pea. Train vines up a trellis to increase blossom density for bees.
Phacelia, Lacey *Phacelia tanacetifolia* **NW:** Southwestern US	M–L–SUC Borage. Curled fiddleheads of light-purple flowers with protruding stamens. Blooms 6–8 weeks. *Plants may irritate skin: wear gloves.* **Group with:** Crimson clover, dill.	Annual Deer resistant.	**2–3 ft (60–90 cm)** Sun to part shade, well-drained, weed-free soil. Can start in pots, but best direct-sown spring, early fall or every 2 weeks. Needs dark for germination; cover seed ¼ in (6 mm). Catches nitrates and prevents them from leaking into groundwater.	BB, HB, SB, BI Top honey plant: 180–1,500 lbs/ac. Light-blue pollen: 300–1000 lbs/ac. Special value to native bees. Stagger bloom time so not blooming at same time as edible crops or may lure bees from them.
Sainfoin, a.k.a. Holy Clover *Onobrychis viciifolia* **OW**	M Legume. Pink pea flowers with purple nectar guides. Gap-filler between dandelion and white sweet clover. May become weedy. Avoid sensitive wild habitat.	Perennial Hardy to zone 7; some cultivars more hardy. Drought resistant.	**30 in (76 cm)** Treat seed with inoculate. Long taproot. Forage for livestock: *Onobrychus* means "devoured by donkeys." Some varieties bloom 3 times a year.	BB, HB, SB Superior honey plant, 10 times more likely to attract honeybees than Dutch clover. Yields nectar in morning. High-quality pollen.
Sunflower *Helianthus annuus* **NW:** Canada except YT, NU, NL; continental US	M Aster. Centre of the flower head a composite of tiny florets. Rays range from creamy white to yellow, orange, red, burgundy and brown.	Self-seeding Annual	**To 13 ft (4 m)** Sun, in medium-moist well-drained soil. Bees cross-pollinate different varieties of sunflowers, often creating something new. Avoid pollen-free cultivars.	BB, HB, SB, BI Honey potential: 30–100 lbs/ac. Pollen: 200–250 lbs/ac. Nectar production varies depending on variety. Specialist pollinators: Sunflower bees *Diadasia* spp. and *Svastra* spp.
Vetch, Hairy *Vicia villosa* **OW**	E–M Legume. Bilabial purple to blue-purple flowers with long corollas. **Group with:** Austrian winter pea, crimson clover, winter rye.	Self-seeding Annual	**3–4 ft (90–120 cm)** Sun, well-drained soil. Seed by mid-August for winter cover crop. (Or October in zone 6.) Nitrogen fixer, green manure, suppresses weeds, improves soil. Easy to pull after blooming. Cattle forage.	BB, HB, SB, BI Heritage honey plant in BC. Honey potential: 22–45 lbs/ac. Exceptional bumblebee plant. Overwintered, can be an important early-spring nectar source.
Vetch, Spring *Vicia sativa* **OW**	E–M Legume. Bilabial purple to blue-purple flowers, long corollas. *Seed toxic to poultry, livestock.*	Annual	**30 in (76 cm)** Similar to hairy vetch but less winter-hardy. Grow with something to climb (like barley or oats) to prevent winter rot.	BB, HB, SB, BI Exceptional bumblebee plant. Important nectar source for long-tongued bees.

*Each pound per acre is equal to approximately 1.1 kilograms per hectare.

Bee Hedgerows

CHAPTER EIGHT

*A*s farming becomes intensified to feed a growing world population, the stress that this places on the bees that pollinate many of our food crops is taking its toll. Eco-agriculturalists are turning to an ancient model of farming that includes growing hedgerows of bee-supportive plants around the borders of fields to help ailing pollinators and bring bees onside to help pollinate crops.

Hedgerows are dense borders of trees, shrubs and vines that can be used to outline fields and contain livestock. Rich with plants bearing fruit and nuts, these hotspots of biodiversity provide food for humans and wildlife, and room and board and blossoms for bees. Layers of blooming trees, shrubs and ground cover act as life-saving corridors of bee refuge, packed with nutrients for adult pollinators and their brood. Ground-nesting bees rely on the undisturbed and sheltered soil under hedgerows to escape the rigorously cultivated earth of annual food crops.

Opposite: Pin cherry flowers growing in Victoria's Spring Ridge Commons, a successful permaculture project that benefits humans, birds and bees. **Right:** Instead of monoculture acres, the more healthy and profitable model looks like a mosaic or patchwork quilt; Dan Jason layers violas, zinnias, lion's tail and sunflowers in his seed-saving garden.

Flowering white strawberries, wild black raspberries and red everbearing raspberries all feed spring bees in our small backyard garden.

A Hedgerow: Deer Resistant and Bee Rich

A successful hedgerow will become an enriched zone for wildlife. Any unfenced bee garden where the deer and the antelope roam will bring in hungry herbivores that can wreak havoc on vegetables, fruits and flowers. Planting a tightly woven hedgerow with plants on the outside that deer can nosh on, and inner layers of shrubs and trees they are not as fond of, will help keep them out of your corn, squash and kale patch. This hedgerow will also protect your tender plants from harsh northern winds and provide habitat for wild songbirds. While no plant is *guaranteed* to repel deer, especially when young and tender, this hedgerow is a mighty good bet. See opposite page for a sample hedgerow garden plan.

Small, organically farmed fields rotated with bee pasture and surrounded by hedgerows have real potential to make our food system better, improving food security for humans and bees. This model can be radically scaled down, too, and integrated into yards big and small, and even into a container garden, where threesomes of small potted shrubs and trees can dole out delicate spring flowers for bees, along with berries to top your morning porridge.

PLANNING YOUR BEE HEDGEROW

A successful bee hedgerow is built on the model of an edible forest, with a succession of blooming layers of fruit-bearing trees and shrubs providing edible fare for people and forage for bees. Hand-to-mouth berry bushes easily accessed by children grow side by side with larger fruit trees, and together they provide a massive biomass of pollen and nectar at crucial times in the life-cycles of many bees. Completing this planting for pollinators are ground covers that thrive in partial shade and suppress weeds around the roots of the lower bushes. Low-growing wintergreen (*Gaultheria procumbens*) and alpine strawberry (*Fragaria vesca*), for example, are understory stalwarts that bloom for bees and also provide jewel-like berries for eating.

In order to plan a successful bee garden, think in threesomes: trios that bloom at the same time or one after the other. A simultaneously blooming bunch should include one or more open-access plants that

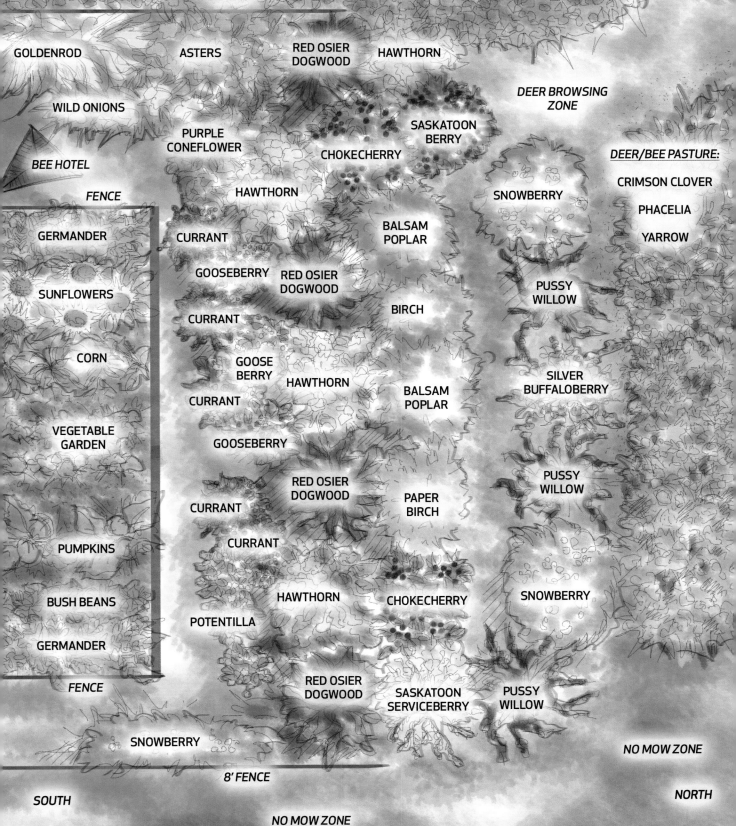

provide nectar and pollen for bees of all stripes. Indian plum (*Oemleria cerasiformis*), blooming earliest, overlapping Oregon grape (*Mahonia aquifolium*) and red-flowering currant (*Ribes sanguineum*), make up one trio of native plants that should be found in numbers in the Pacific Northwest to support bees at the critical time of the year just before domestic fruit trees are at their peak blossom time. Salmonberries (*Rubus spectabilis*), followed by wild red raspberries (*Rubus idaeus*) and finally native blackberries (*Rubus allegheniensis),* blooming from early to mid- to late spring, provide a life raft of continuous bloom for bees.

HOW TO PLANT SHRUBS AND TREES

Plant trees and shrubs when they are dormant, in early spring or late fall. First, though, consider how the tree

Opposite, clockwise from top left: Blackberry flowers provide food for bees when many other fruit-bearing shrubs have finished blooming. Snowberries are the backbone of a successful bee hedgerow. The hawthorn tree is an essential part of European hedgerows; in the New World they bloom at a crucial time for many species of bees active in spring. Jewel-like currants can be used as edible garnish on sweet puddings.

might shade other plants in your garden and change the microclimate in your yard. You may not want a tree that blocks sunlight from your food garden or perennial border. Varieties grafted onto dwarf root stock are often the best choice for a smaller space.

1. **SELECT A SITE:** Make sure it will accommodate the fully grown size of your shrub or tree. Imagine the drip line when it is fully grown; that's how far the roots will reach out. While trees need space around the roots and won't do well with other trees or shrubs within this comfort zone, many are fine with an overplanting of shallow-rooted ground cover or bulbs, building up blossom density for bees.

2. **DO YOUR RESEARCH:** Find out the potential spread and height of the shrub or tree, and whether or not the shrub sends up shoots, requiring more maintenance, as raspberries do. If you are planting in soils heavy in clay or soppy soil, choose trees and shrubs suited to this habitat (see Trees for Bees, page 169). Native plants are good for bees and do best in native soil, which saves you the extra work of amending it.

3. **PICK YOUR PLANT:** Avoid buying root-bound shrubs and trees that can become impervious to moisture. Choose plants from organic nurseries, avoiding those that use pesticides and herbicides on the land around the stock, if not on the shrubs or trees themselves.

4. **PREPARE THE SOIL:** Shrubs and trees need to be able to get their roots into the soil to drink up moisture and minerals. This means you need to prepare the soil by creating an area of friable (easily crumbled) earth that goes beyond the size of the pot and the plant's current root ball.

5. **DIG THE PLANTING HOLE:** If the soil you are planting into is extremely dry, add a couple of inches of water to the bottom of the hole to give the plant a good start. Rough up the sides of the hole so the

---

Poor Drainage?
Try Double Digging

I learned how to double dig from Catherine Shapiro, an experienced permaculture gardener. Prepare a bed by removing a layer of soil—about 1 to 2 feet (30 to 60 cm) in depth. The depth and width depends on what you're planting and the length and spread of its roots. Put the soil to one side, and get a pitchfork in to loosen the soil in the hole to the depth of its tines. Mix equal parts sand and compost into the soil you removed before replacing it, to make up about one-third of the total soil volume.

---

roots can work their way sideways as well as down. The hole should be no deeper than the roots, but three times the diameter of the pot. If the water sits on the bottom of the hole, you'll know the drainage is poor. If your soil is heavy or compacted, consider double-digging (see the Poor Drainage? Try Double Digging sidebar above).

6. **PREPARE THE PLANT:** Water the plant thoroughly in the pot. Gently tip the pot over and grasp the tree or shrub by the main stem or trunk to pull the plant out. After you've removed it from the pot, massage the roots to gently tease them out. If they are still dry, soak them for a few minutes in water.

7. **PLACE THE PLANT:** Place the shrub or tree in the hole. If the plant is spindly it may need staking— use soft ties to attach it to the support. Refill the hole, making sure the soil gets in between the roots so there are no air pockets, and firmly pat down the soil. Don't cover the stem beyond where it was

If you put freshly cut pussy willows in water, they will produce pollen, which is a great way to teach children about this important food for spring bees.

mining bees. Pussy willows are wind-pollinated, but produce essential pollen for spring bee broods. High-maintenance plants, they are not well-suited to urban backyards because their extensive root systems clog drains. But, for this reason, they are great for stabilizing soil in riparian environments and steep slopes. Back to the home front, weeping goat willows (*Salix caprea* 'Pendula') can be grafted onto yard-friendly tree stock as an ornamental feature in front gardens. They also attach well to dwarf stock, which means you can even grow pussy willows in pots.

submerged in soil in its pot. Mulch 1 to 2 inches (2.5 to 5 cm) high with leaf litter, shredded bark or compost, a good hand's length away from the trunk.

8. **MAINTAIN THE PLANT:** Keep a diameter of about 3 feet (0.9 m) weed-free around the shrub or tree for the first 3 years. Water around the roots to establish it and then once or twice a week as needed. Do not over-prune, but do your research to devise a snipping strategy.

SPOTLIGHT ON PUSSY WILLOW: *THE* BEST SHRUB FOR EARLY-SPRING BEES

Some of my fondest childhood memories involve searching for pussy willows (*Salix discolor*) to take home for Easter bouquets. At my mom's urging, we would load into the blue Chevy and dad would drive us wildly through farmers' summerfallow fields to a slough surrounded by willows, where we would search for the plumpest fuzzy catkins. The list of bees and other insects supported by these willows is long and diverse, making them powerhouse plants for spring microfauna, such as blue orchard bees and *Andrena*

SPOTLIGHT ON ROSES: A SCENTED TAPESTRY FOR BEES

During World War II, girl guides in the United Kingdom collected sacks of rosehips from hedgerows to make into a vitamin-C syrup to help prevent scurvy among the masses. Wartime mothers made them into preserves, jellies and stewed compote, and even tossed hulled rosehips into baking like raisins. The best roses for bees are wild, bushy, climbing and tumbling; tightly wound tea roses may be elegant, but are con artists and time-wasters to bees. Roses have got to be simple, with beautiful fringes of stamens loaded with pollen that little bees and bumblebees can swim inside, coating their bodies with a rich blanket of scented powder. Roses must have scent, filling the air on a hot, muggy day with seductive perfume. And a hedgerow gives you the opportunity to grow a tapestry of heritage rose shrubs—including the native varieties that are best for indigenous bees. If you have the support structure, an archway covered in climbing roses can boost the pollen and nectar power of your garden in a big way with bees singing above and beyond in summer choruses.

Opposite, clockwise from top left: My mom taught me how to extract the tart juice from crabapples. Red flowering currants provide important sustenance for bumblebee queens. Rosehips infused in wine make a delicious tonic. Some species of native roses benefit from buzz pollination.. **Above:** A mountain ash tree grows next to a beehive at UBC Farm.

PRAIRIE SUPERFOODS: SASKATOON AND SILVER BUFFALOBERRY WITH WILD ROSE

Silver buffaloberry, saskatoon berry and wild rose are winter-hardy, fruit-bearing shrubs with blossoms accessible to a wide variety of bees. Silver buffaloberry (*Shepherdia argentea*) is a superfood for bees, birds, bears and humans. A traditional ingredient in pemmican, the red-orange or yellow fruits are loaded with antioxidants. The blossoms of Saskatoon berries (*Amelanchier alnifolia*) feed guilds of prairie bees with an early source of pollen and nectar, and wild stands still bring folks out with buckets to gather fruit for fowl supper pies. Silver buffaloberry and Saskatoons are technically shrubs but can grow to the height of small trees. They benefit from the layering-in of lower

Planting Bee Hedgerows Makes Good Financial Sense

You may love blueberries, but if you plant only blueberries in your yard, the native bees will make a run for ecologically greener and more biodiverse pastures. Brian Campbell tells a sad story about a BC blueberry farm: the original owner had worked on a sustainable model, planting bee pasture and hedgerows to attract pollinators. When a new owner bought the farm, the bank based his mortgage on his ripping out the hedgerows and natural habitat to plant every single square foot with blueberries. When his blueberries blossomed, he had a pollinator deficit—the first this farm had ever faced. What to do? He had to spend money renting honeybee hives to replace the pollinators supported by the hedgerows, money that would not have been lost if he had kept this pollinator habitat.

shrubs to support each other in harsh weather, weaving together a peaceful refuge for bees. A later-blooming low shrub to add to this trio is the classic wild prairie rose (*Rosa acicularis*), an essential aromatic plant for bees. These three shrubs meld into a dense habitat that protects nesting birds and provides habitat for ground-nesting bees. Add buckbrush (*Symphoricarpos occidentalis*), crabapples (*Malus* spp.), shrubby cinquefoil (*Dasiphora fruticosa*) and chokecherries (*Prunus virginiana*) for the ultimate winter-hardy hedgerow of edibles.

A WEST COAST TRIO: OCEANSPRAY, OREGON GRAPE AND BLUEBERRY

In my neighbour Celia's boulevard, blue orchard mason bees zip back and forth from nest holes in the power pole to the fragrant yellow blossoms of Oregon grape (*Mahonia aquifolium*). The low, medium and high varieties layer well into a hedgerow, providing good "bones" to build upon. Their berries can be mixed with other bush fruit, such as blueberries, to make jams and jellies, or a tart infusion to satisfy thirst in early-summer days. Blueberry bushes are also available in varying heights, blooming slightly later to provide continuing nectar and some pollen for buzz-pollinating bees. Finally, oceanspray (*Holodiscus discolor*) can provide late-spring forage for bees of all stripes, especially as it grows to towering heights. On sunny days in June, the foamy white flowers churn with insect activity.

Berries Need Bees

Each sun-warmed blueberry you pop into your mouth was potentially visited by at least 2 bumblebees, 2 mason bees, or 3 to 10 honeybees, depositing about 125 grains of pollen.[53] If there are no bees at all and a pollination deficit, the berries will be misshapen and inedible, or there won't be any fruit at all. So let's do the math: an average cup of 220 berries requires 440 bumblebee and/or mason-bee visits, 660 to 2,200 honeybee visits, or orchestras of bees working together in concerts of pollination diversity.

Below: Spirit Hills Winery makes mead infused with wild Alberta rose petals. **Opposite, clockwise from top left:** Silver lindens are reported to be toxic to bees. Apple and crabapple provide prolific blossoms in spring for bees. If you are tall enough to reach the blossoms of the tulip tree, you can drink the nectar. Mason bees love the blossoms of Oregon grape.

Although the fragrant blossoms have been used for traditional medicine, it is not advised to consume any part of ocean spray.

A TEA AND TONIC MIX: WILD BLACK CHERRY, NEW JERSEY TEA AND EASTERN TEA

Plants in the *Ceanothus* genus are important for bees, and New Jersey tea (*Ceanothus americanus*) is a great example, supporting a broad spectrum of bees and other insects. Like scarlet bee balm, (*Monarda didyma*; see page 82 for planting advice), it was used to make a beverage for tea-deprived settlers during the American Revolution. Wintergreen, a.k.a. eastern tea berry (*Gaultheria procumbens*), is a hardy buzz-pollinated understory plant with berries, twigs and leaves that can also be plucked for teas and tonics. And in spring, the branches of wild black cherry (*Prunus serotina*) are densely packed with shallow white blossoms, making it one of the best trees for bees—and for early settlers, too, who added the fruit to rum for a kick called "cherry bounce." As a final touch, a carpet of bunchberry (*Cornus canadensis*), with edible fruits tasting like apples, added as ground cover will provide a blossom-rich sun-warmed tonic for thirsty bees.

FLOWERING TREES: VERTICAL BEE GARDENS

There are days I wish that I could be like Alice in Wonderland, drinking a potion that makes me grow to a great height to watch the bees in the sunny tops of linden and cherry trees. Trees are indispensable for bees, with the potential to produce a large amount of nectar and pollen in a vertical space. As land use becomes more densely developed and farmed, trees play an increasingly important role in the provision of nectar and pollen for bees. One American tulip tree

---

Nurse Trees for Hedgerows

Hedgerows are planted with a mix of nurse trees and slower-growing trees. Nurse trees are fast-growing trees that help the slower trees and shrubs by providing shelter, protection and support and sometimes improving soil health. Nurse trees include balsam poplar (*Populus balsamifera*), pin cherry (*Prunus pensylvanica*), aspen (*Populus* spp.) and Jack pine (*Pinus banksiana*). Poplar and pine can provide resin for honeybees and resin bees, and pin cherry trees provide pollen and nectar for a wide variety of bee species. Aspen trees are used by male bumblebees to leave their scent markers for mating. Four fantastic nurse trees that feed bees and fix nitrogen are from the legume family: black locust (*Robinia pseudoacacia*), honey locust (*Gleditsia triacanthos*), eastern redbud (*Cercis canadensis*) and golden chain tree (*Laburnum* spp.). As nurse trees die off, they can be left to decompose and provide homes for bees while enriching the soil for the remaining plants.

(*Liriodendron tulipifera*), for example, can produce an incredible 4 cups (1 L) of nectar, enough to produce 2.2 pounds (1 kg) of honey. It is estimated that 5 or 6 large trees densely laden with blossoms can provide more nectar and pollen than an acre of bee meadow.[54] Trees also provide nesting sites, oils and antimicrobial resins that honeybees use to make propolis.

TREES
for Bees
A SEASONAL PLANTING CHART

*T*rees can provide a large volume of pollen and nectar, but many only bloom for two weeks. It is essential to plan your garden to stagger the bloom time of trees to provide an optimum calendar of forage for bees. Planting companions that bloom at different times from your fruit and nut trees is also key, so that bees are not lured away at that crucial time when they are needed most. It's worth doing a bit of research on pruning for each species, but do not be tempted to over-prune, and leave large trees to professional arborists.

PLANT ORIGIN
OW: Old World (indigenous to Europe, Asia or Africa)
NW: New World (indigenous to the Americas and their islands)

BLOOM PERIOD
E: Early
M: Mid-season
L: Late
SUC: May be seeded in succession

HARDINESS
refers to the coldest zone the plant is hardy to.
DEER AND DROUGHT RESISTANCE are most reliable once the plant is established.

BEES ATTRACTED
BB: Bumblebees
HB: Honeybees
SB: Solitary bees
BI: Beneficial insects

PLANT AND NATIVE REGIONS	PLANT FAMILY, BLOOM, FRUIT AND COMPANIONS	HARDINESS	HEIGHT AND GARDEN NOTES	BENEFITS TO BEES*
Apple, Domestic *Malus domestica* **OW**	E Rose. Shallow cup-shaped white or pink blossoms. Edible fruit. **Group with:** 2 other apples or crabapples with similar bloom time for pollination.	**Perennial** Hardy to zone 2–5, depending on variety.	**10–30 ft (3–9 m)** Sun in average to fertile, slightly acidic, well-drained soil. Prune branches in winter, shoots in summer. Can be espaliered against trellises or fences. Average to moist conditions.	BB, HB, SB Honey potential: up to 45 lbs/ac. Pollen collected in morning. Digger, mason, mining, sweat bees. Later-blooming varieties help fill June nectar gap.
Arbutus, a.k.a. Pacific Madrona *Arbutus menziesii* **NW:** Pacific coast	E Heather. Bell-shaped blossoms. Edible orange berries enjoyed by birds. **Group with:** Oregon grape, red osier dogwood, snowberry.	**Evergreen Perennial** Hardy to zone 2.	**30–50 ft (9–15 m)** Sun in sandy, rocky, well-drained soil (as similar to native soil as possible). Do not overwater or fertilize. Prune late winter or early spring.	BB, HB, SB, BI Honey plant. Mason bees. Significant bumblebee, butterfly and hummingbird plant.
Ash, Wafer, a.k.a. Hop Tree *Ptelea trifoliata* **NW:** ON, most of US	E Citrus. Clusters of fragrant shallow flowers with white to yellow-green petals. Revered by First Nations as a sacred medicinal tree. **Underplant:** Long-headed coneflower.	**Perennial** Hardy to zone 3. Deer resistant.	**20 ft (6 m)** Sun to light shade in well-drained soil, medium to moist. Prefers rich but will tolerate poor soil if moist. Prune suckers. Food plant for butterfly larvae. Endangered in Ontario.	HB, SB Nectar contains a medicinal chemical for bees (scopolamine N–oxide) that helps fine-tune flight muscles and keeps bees calm.[55]
Buckeye, Ohio *Aesculus glabra* **NW:** ON, mid-eastern US	E Soapberry. Yellow-green blossoms, protruding stamens. See horsechestnut. Plant instead of California buckeye (*A. × carnea*), which is toxic to honeybees. Toxic seeds/tree.	**Perennial** Hardy to zone 4. Road-salt tolerant. Deer resistant.	**30–50 ft (9–15 m)** Partial shade in deep, moist, well-drained loam. Prune in late fall or winter. Bark gives off fetid odour when damaged.	BB, HB, SB, BI Minor honey plant. Digger, long-horned and mason bees. Hummingbird and butterfly plant.
Catalpa, Northern *Catalpa speciosa* **NW:** Mississippi Delta	E–M Bignonia. Showy white orchid-like fragrant flowers with dark-purple nectar guides. Allelopathic so large groups may inhibit growth of some plants. Invasive east of native habitat. **Underplant:** Clematis, indigo.	**Perennial** Hardy to zone 5.	**50–100 ft (15–30 m)** Sun to part shade, well-drained but moist, fertile soil. Trim suckers. Prune damage in summer. Good adapter to environmental change. Suitable for bee pasture.	BB, HB, SB Honey tree with extra floral nectaries. High-protein pollen. Flowers help fill June gap. Large carpenter bees. Honeybees gather purple resin for propolis.
Catalpa, Southern *Catalpa bignonioides* **NW:** Southeastern US	E–M Bignonia. See northern catalpa. The worms from the catalpa moth have a symbiotic relationship with this tree. Do not eradicate them.	**Perennial** Hardy to zone 5.	**30–40 ft (9–12 m)** See northern catalpa. Enjoy a fragrant picnic under its boughs while the bees hum above you.	BB, HB, SB, BI See northern catalpa. Significant bumblebee and honeybee plant.
Cherry, Pin *Prunus pensylvanica* **NW:** Canada except YT; northern US	E Rose. White blossoms. Edible sour fruit for jelly and wine. **Underplant:** Aster, borage, bunchberry, strawberry, thyme, violet, zinnia. See wild black cherry.	**Perennial** Hardy to zone 1.	**20–40 ft (6–12 m)** Sun to light shade, moderate well-drained soil. Prune in August if necessary. Fruit is exceptional food for wildlife.	BB, HB, SB, BI Mason, mining, sweat bees and more. Extra floral nectaries. Butterfly, moth plant. Higher blossom density than sweet cherries.
Cherry, Sweet *Prunus avium* **OW**	E Rose. Shallow pink or white blossoms. Edible red fruit. **Group with:** Compatible varieties for pollination. Avoid plum. See pin cherry.	**Perennial** Hardy to zone 4.	**20–50 ft (6–15 m)** Sun, well-drained soil. Prune in August if necessary. Avoid ornamental cherries with double or frilly blossoms.	BB, HB, SB Honey potential: 22–45 lbs/ac. Sugar concentration 28 percent. Pollen high in protein. Mason bees.

*Each pound per acre is equal to approximately 1.1 kilograms per hectare.

PLANT AND NATIVE REGIONS	PLANT FAMILY, BLOOM, FRUIT AND COMPANIONS	HARDINESS	HEIGHT AND GARDEN NOTES	BENEFITS TO BEES*
Cherry, Wild Black *Prunus serotina* **NW:** BC, ON, QC, NB, NS; central, eastern, southern US	E Rose. Clusters of shallow white blossoms. Edible sour fruit for jelly and wine. The rest of the tree is toxic. **Underplant:** Bunchberry, strawberry, violet.	Perennial Hardy to zone 4. Deer and drought resistant.	Up to 80 ft (24 m) Sun to light shade, moderate, well-drained soil. Prune in August if necessary. Fast-growing. Fruits smaller than sweet cherries, exceptional food for wildlife.	BB, HB, SB, BI Mason, mining, sweat bees and more. Extra floral nectaries. Significant butterfly and moth plant. Higher blossom density than sweet cherries.
Crabapple, American *Malus coronaria* **NW:** East of Mississippi	E Rose. Pink blooms. High-pectin fruit for delicious jelly. See domestic apple. **Underplant:** Allium.	Perennial Hardy to zone 4.	10–30 ft (3–9 m) See domestic apple, but prefers even moister environment. Fruit is key food for wildlife.	BB, HB, SB Wild crabapples are an important source of nectar for native bees. See domestic apple.
Crabapple, Pacific *Malus fusca* **NW:** Pacific coast	E Rose. See American crabapple. Flowers and fruits produced on spurs.	Perennial Hardy to zone 7.	30 ft (9 m) See domestic apple. Flowers and fruits support a wide variety of wildlife.	BB, HB, SB See American crabapple. Significant plant for blue orchard mason bees. Visited by hummingbirds.
Hawthorn, Black *Crataegus douglasii* **NW:** Pacific Northwest	E Rose. Masses of shallow white blossoms. Thorns. Small, seedy edible fruit. Loved by birds for food and nesting habitat. **Group with:** Bee balm, New Jersey tea, saskatoon berry.	Perennial Hardy to zone 4. Drought resistant.	20–46 ft (6–14 m) Sun to partial shade, dry to moist clay, sand and loam. Prune late winter/early spring. Ash on soil surface deters disease. Boulevard tree.	BB, HB, SB, BI Honey potential: 22–45 lbs/ac. Essential bee hedgerow plant loved by many bee and butterfly species.
Hawthorn, Cockspur *Crataegus crus-galli* **NW:** ON, QC, NS, eastern US	E Rose. See black hawthorn.	Perennial Hardy to zone 5. Drought resistant.	10–15 ft (3–4.6 m) See black hawthorn. Tolerant of heat and salt spray from cars. Thornless cultivar is 'Inermis'.	BB, HB, SB, BI Rich supply of cream-coloured pollen at the peak of brood-rearing season for honeybees. See black hawthorn.
Hazel *Corylus americana* **NW:** MB, ON, QC, eastern US north of FL	E Birch. Long green catkins similar to many wind-pollinated trees. Edible nuts. **Group with:** Alder and willow, other important early sources of pollen.	Perennial Hardy to zone 3.	10–30 ft (3–9 m) Sun to partial shade in fertile, well-drained soil. Choose disease-resistant cultivar. Prune late fall to early spring. Among the wind-pollinated trees bees harvest pollen from.	HB Pollen: 180–445 lbs/ac. Light-green pollen if late winter is warm while in bloom. Key for brood-building for honeybees and bumblebees.
Horsechestnut *Aesculus hippocastanum* **OW**	E Soapberry. White or pink blossoms with yellow blotch. Toxic seeds. Allelopathic. Toxic seeds/tree. **Underplant:** Virginia bluebells.	Perennial Hardy to zone 2. Pollution tolerant.	30–60 ft (9–18 m) Partial shade in deep, moist, well-drained loam. Prune in late winter if necessary. Keep watered.	BB, HB, SB Honey tree. Long-tongued bees. Yellow nectar guides turn red when pollinated. Blooms one month.
Japanese Snowbell *Styrax japonicus* **OW**	E Styrax. Pendulous white blossoms with protruding stamens. **Underplant:** Virginia bluebells.	Perennial Hardy to zone 5. Deer resistant.	20–30 ft (6–9 m) Sun to partial shade in moist, acidic soil. Prune in late winter or very early spring.	BB, HB, SB Pollen and nectar. A good tree to help fill the June nectar gap.
Linden, a.k.a. Basswood *Tilia americana* **NW:** Central and eastern Canada and US	E–M Mallow. Clusters of highly fragrant pale-yellow blossoms. Plant several linden varieties to extend bloom. **Underplant:** Lupin.	Perennial Hardy to zone 2.	60–120 ft (18–37 m) Sun to partial shade in deep, moist, well-drained loam. Plant in fall. Prune damaged limbs in winter. Blossoms can be dried for tea to treat colds, migraines.	BB, HB, SB, BI Honey potential: 90–445 lbs/ac. Sugar concentration up to 76 percent. Sweat bees.

PLANT AND NATIVE REGIONS	PLANT FAMILY, BLOOM, FRUIT AND COMPANIONS	HARDINESS	HEIGHT AND GARDEN NOTES	BENEFITS TO BEES*
Locust, Black *Robinia pseudoacacia* NW: Appalachia	E Legume. Pea-like bunches of white flowers. Most of this plant is toxic, including seeds. Invasive in ON, QC, NS. **Underplant:** Bee balm, Joe-Pye weed, New Jersey tea, saskatoon berry.	Perennial Hardy to zone 4. Deer resistant. Tolerates black walnut.	40–80 ft (12–24 m) Sun, dry to moist conditions in loamy soil. Grows well in poor, sandy soil. Prune hard in spring or any time apart from blooming. Pioneer plant useful for ecological restoration.	BB, HB, BI Honey potential: 90 to more than 180 lbs/ac. Key bumblebee tree. Sugar concentration 34–67 percent. Low-protein pollen.
Locust, Honey *Gleditsia triacanthos* NW: ON, Mississippi Delta	E–M Legume. Drooping clusters of fragrant yellow-green flowers with protruding stamens. Thorns. Edible pods used as famine foods in the 1930s and farmed as livestock feed. Can spread aggressively. **Underplant:** Lupin.	Perennial Hardy to zone 4. Deer resistant. Pollution tolerant.	75–100 ft (23–30 m) Sun to part shade, medium to moist. Prune after blooming during dry periods. Pioneer species, suitable for habitat restoration and soil stabilization.	HB, SB, BI Honey potential 45–90 lbs/ac. Excellent bee pasture. Mining and sweat bees. Significant bumblebee plant. Flowers can help fill June nectar gap.
Maple, Red Swamp *Acer rubrum* NW: Eastern Canada and US	E Maple. One of the first trees to flower in spring. Male and female flowers, sometimes on separate trees. **Underplant:** Dutchman's breeches.	Perennial Hardy to zone 3.	160 ft (49 m) or higher Sun to light shade in deep, moist, well-drained, slightly acidic soil. Prune in fall only if necessary. Heritage honey plant in eastern Canada.	HB, SB Honey plant in warm spring: (*Acer* spp.) to 445 lbs/ac. Sugar concentration up to 52 percent. Mining bees.
Maple, Vine *Acer circinatum* NW: Pacific coast	E Maple. Red flowers, protruding stamens. **Group with:** Oceanspray.	Perennial Hardy to zone 6. Drought resistant.	10 ft (3 m) Part shade to shade, moist soil. Prune to shape in midsummer. Starts as shrub and can become a tree. Understory to tall evergreens.	HB, SB Heritage honey plant in BC. Mining bees. Significant plant for mason and bumblebees.
Mountain Ash, 'Pink Pagoda' *Sorbus hupehensis* 'Pink Pagoda' OW	E–M Rose. Clusters (panicles) of small, white, shallow flowers. Pink berries. **Underplant:** Crocus.	Perennial Hardy to zone 4.	20–23 ft (6–7 m) See Swedish whitebeam. Gorgeous berries feed wildlife in winter.	BB, HB, SB, BI See Swedish whitebeam.
Peach and Nectarine *Prunus persica* OW	E Rose. Shallow white to pink flowers. Delicious edible fruits. **Underplant:** Native strawberry, wild onion.	Perennial Hardy to zone 6.	15–26 ft (4.6–8 m) Sun and moderately drained, fertile soil. Prune in late winter. Suitable for espalier. Dwarf varieties can be grown in pots.	BB, HB, SB Honey potential: (*Prunus* spp.) to 45 lbs/ac. High-protein peach pollen. Cuckoo, mining, small carpenter, sweat bees.
Pear *Pyrus communis* OW	E Rose. Shallow white cup-shaped bloom. For pollination, may need another variety. See peach and nectarine.	Perennial Hardy to zone 6.	26–30 ft (8–9 m) Sun and moderately drained, fertile soil. Prune 20 percent of last year's growth in winter.	BB, HB, SB Honey tree. Red-yellow pollen and nectar. Food for spring bee brood.
Plum *Prunus domestica* OW	E Rose. Shallow white to pink bloom. Delicious fruit. Requires pollenizer tree. Toxic seeds. **Underplant:** Chives, mint.	Perennial Hardy to zone 4.	10–30 ft (3–9 m) See pear. Does well in soils heavy in clay or limestone.	BB, HB, SB See pear. Traditional honeybee plant.
Plum, Canada *Prunus nigra* NW: MB, ON, QC, northeastern US	E Rose. Shallow white to pink bloom. Toxic seeds. **Underplant:** bee balm, chives, mountain mint, strawberry.	Perennial Hardy to zone 3.	30 ft (9 m) See peach and nectarine. Suitable for boulevard trees.	BB, HB, SB Cuckoo, mining, small carpenter and sweat bees. Special value to native bees.

On warm days in late winter, honeybees collect hazel pollen for their brood. Since the catkins are wind-pollinated, honeybees find the pollen difficult to collect because it is not sticky.

PLANT AND NATIVE REGIONS	PLANT FAMILY, BLOOM, FRUIT AND COMPANIONS	HARDINESS	HEIGHT AND GARDEN NOTES	BENEFITS TO BEES*
Poplar, Balsam *Populus balsamifera* **NW:** Canada, most of US	E Willow. Long green catkins. Avoid planting near green alder. **Group with:** Chokecherry, Indian plum, ninebark, red osier dogwood.	**Perennial** Hardy to zone 1.	**75–100 ft (23–30 m)** Sun in moist soil. Prune damaged branches in late spring or early summer. Wind-pollinated. Sap collected to make healing "balm of Gilead."	HB Nectar and high-protein pollen. Sap collected by bees to make propolis.
Pussy Willow *Salix discolor* **NW:** Canada except YU and NU; northern continental US, ID to ME	E Willow. Fuzzy grey catkins. Male flowers produce pollen and nectar. Females grow on separate trees and produce only nectar. **Group with:** buttonbush, catalpa, elderberry.	**Perennial** Hardy to zone 2.	**150 ft (46 m)** Sun, evenly moist soil. Willows stabilize soil but roots can damage drainage systems and pipes. Suitable for pond edge, steep banks. Pioneer plant useful for ecological restoration. Prune after flowers fade.	BB, HB, SB, BI Best-quality pollen for spring brood: Up to 1,340 lbs/ac. Considered by many beekeepers to be *the* most important honeybee plant. Cuckoo, large and small carpenter, mason, mining, sweat bees.
Redbud, Eastern *Cercis canadensis* **NW:** ON, southeastern US	E Legume. Fragrant, pink, pea-shaped flowers are edible and can be eaten in salads. **Underplant:** Indigo, lupin.	**Perennial** Hardy to zone 5.	**20–30 ft (6–9 m)** Partial sun, mix of clay loam and rocky soil in medium to moist conditions. Prune in spring after blooming. Protect from strong wind.	BB, HB, SB Honey tree. Cuckoo, digger, long-horned, mason, mining and sweat bees.
Tulip Tree, American *Liriodendron tulipifera* **NW:** ON, eastern US	E–M Magnolia. Large tulip-shaped yellow-green blossoms. These thirsty trees do best without understory or competition from other roots.	**Perennial** Hardy to zone 5.	**160 ft (49 m) or higher** Full to partial sun in fertile soil. Prune in winter if necessary. Largest native tree in eastern North America. Suitable for large yard. Significant food and shelter plant for wildlife.	BB, HB, SB Major honey plant. Long-tongued bees. Each blossom can produce up to 1 tsp (5 mL) nectar, a tree up to 9 lbs (4 kg) of nectar.
Whitebeam, Swedish *Sorbus intermedia* **OW**	E–M Rose. Clusters of small, white, shallow flowers. Orange to red berries. Spectacular fall foliage. The "dogberries" are made into jam.	**Perennial** Hardy to zone 4.	**82 ft (25 m)** Full to partial sun, somewhat fertile, acidic, well-drained soil, moist to moderate. Prune to shape in late winter or early spring. Berries loved by birds.	BB, HB, SB, BI Nectar. Many *Sorbus* are pollinated by bees, flies and other insects. Blossoms can give off a slightly fishy aroma.

*Each pound per acre is equal to approximately 1.1 kilograms per hectare.

SHRUBS for *Bees*

A SEASONAL PLANTING CHART

*F*or your hedgerow, edible forest or other plantings, look for shrubs native to your region and/or disease-resistant cultivars. In-depth research on pruning will help fruit-bearing shrubs feed you and the bees. Choose a selection that bloom at different times to ensure a succession of bee forage in your garden. Add layers of plants blooming at different heights to increase biodiversity of flora and fauna.

PLANT ORIGIN

OW: Old World (indigenous to Europe, Asia or Africa)

NW: New World (indigenous to the Americas and their islands)

BLOOM PERIOD

E: Early

M: Mid-season

L: Late

SUC: May be seeded in succession

HARDINESS

refers to the coldest zone the plant is hardy to.

DEER AND DROUGHT RESISTANCE are most reliable once the plant is established.

BEES SUPPORTED

BB: Bumblebees

HB: Honeybees

SB: Solitary bees

BI: Beneficial insects

PLANT AND NATIVE REGIONS	PLANT FAMILY, BLOOM, FRUIT AND COMPANIONS	HARDINESS	MAXIMUM HEIGHT × SPREAD AND GARDEN NOTES	BENEFITS TO BEES*
Blackberry, Common *Rubus allegheniensis* **NW:** BC, ON, QC, Maritimes, central US, CA	E–M Rose. Shallow flowers, 5 elongated white petals. Edible berries. **Group with:** Native roses, salmonberry.	**Perennial** Hardy to zone 3.	**8 ft (2.4 m) × 8 ft (2.4 m)** Sun to partial shade, humus-rich soil, forest-floor conditions. Prune canes after fruiting. Suitable for understory, edible forest.	**BB, HB, SB, BI** Honey potential: 4–65 lbs/ac. Cuckoo, leafcutter, mason, mining, sweat bees. Nectar/grey pollen.
Bluebeard *Caryopteris × clandonensis* **OW**	M–L Mint. Clusters of small purple or blue flowers, long protruding stamens. **Group with:** Aster, goldenrod.	**Short-lived Perennial** Hardy to zone 5. Deer and drought resistant.	**4 ft (1.2 m) × 4 ft (1.2 m)** Sun to partial shade in well-drained, sandy soil. Roots need winter-protective mulch. Cut back new growth 1 in (2.5 cm) in spring; dress with compost.	**BB, SB, BI** Good late-season source of pollen and nectar. Large carpenter, plasterer, sweat, wool carder bees. Butterfly plant.
Blueberry, Black, a.k.a. Black Huckleberry *Vaccinium membranaceum* **NW:** NWT, YT, NU, BC, AB, ON, south to UT, northern CA	E Heather. Pink buds and white bell-shaped blossoms. Fruit more nutritious and delicious than cultivated blueberries. Beautiful fall foliage. **Group with:** Evergreen huckleberry, salal.	**Perennial** Hardy to zone 5. Drought resistant.	**5 ft (1.5 m) × 5 ft (1.5 m)** Sun to partial shade, rich acidic soil, spreads in moist conditions. Requires air circulation and drainage. Plant several shrubs (not just from cuttings of a single plant) to ensure cross-pollination. Little pruning required. Suitable for raised beds. Heavy feeder.	**BB, HB, SB** Up to 38 species of native bees visit blueberries, including mining bees. Nectar robbed by short-tongued bumblebees. Buzz-pollinated.
Blueberry, Lowbush, a.k.a. Wild Blueberry *Vaccinium angustifolium* **NW:** NU, MB, ON, QC, Maritimes, northeastern US	E Heather. Pink buds and white bell-shaped blossoms. Remove first-year blooms to divert energy to roots. Nutritious edible berries. **Group with:** Evergreen huckleberry, salal.	**Perennial** Hardy to zone 2.	**2 ft (60 cm) × 2 ft (60 cm)** Sun to light shade in acidic, well-drained, sandy soil rich in organic matter, moist to moderately dry. Must prune once established; will not flower following year so alternately prune 2 bushes.	**BB, HB, SB** See black blueberry. Buzz pollinated primarily by bumblebees and mining bees. Specialist pollinator: Maine blueberry bee *Osmia atriventris*.
Buckbrush, a.k.a. Western Snowberry *Symphoricarpos occidentalis* **NW:** NU, BC, Prairies; northern and central US	M Honeysuckle. Shallow white to pink flowers, protruding stamens. Toxic white berries contain saponin. **Group with:** Blanket flower, prairie smoke.	**Perennial** Hardy to zone 2.	**40 in (1 m) × indefinite** Sun, well-drained soil. Tolerates poor soil. Prune in late winter or early spring. Suitable for prairie meadow as well as hedgerow. Easily spread by rooting suckers.	**BB, HB, SB** Mostly nectar, some pollen. Digger, long-horned, mason, mining, small carpenter, sweat, plasterer bees.
Bunchberry *Cornus canadensis* **OW/NW:** Canada, northern US, Rocky Mountains	E–M–L Dogwood. Flowers have 4 white bracts with multiple florets in the middle. Edible medicinal berries high in pectin. **Group with:** Blue-eyed grass, violet.	**Perennial** Hardy to zone 2.	**8 in (20 cm) × 3 ft (0.9 m) per yr.** Partial sun to shade in rich, acidic soil. Difficult to establish; add decayed wood chips, pine needles to soil. Cut back dead branches early spring, otherwise resist pruning. Good for bogs, understory for pine.	**BB, HB, SB, BI** Nectar and pollen. Cuckoo, mining and sweat bees. Petals have mechanism that shoots pollen into the air onto the hairy bodies of bees.
Buttonbush *Cephalanthus occidentalis* **NW:** ON, QC, NB, NS, eastern US, CA	M Bedstraw. Globes of small white flowers, protruding stamens. **Group with:** Pussy willow, swamp milkweed.	**Perennial** Hardy to zone 4.	**20 ft (6 m) × 10 ft (3 m)** Partial to full shade in wet, swampy conditions. Prune in early spring if needed. Suitable for pond edge.	**BB, HB, SB, BI** Honey plant. Cuckoo, green sweat, leafcutter, long-horned bees.
California Lilac *Ceanothus × delileanus* 'Gloire de versailles' **NW:** CA	M–L Buckthorn. Masses of tiny light-purple flowers. **Group with:** Oceanspray, snowberry.	**Perennial** Hardy to zone 6. Drought resistant.	**5 ft (1.5 m) × 5 ft (1.5 m)** Sunny, sheltered location with well-drained soil. Prune after blooming. Water moderately to establish, but do not overwater in subsequent years. Mulch for winter protection.	**BB, HB, SB** Digger leafcutter, small carpenter bees; mason, mining, sweat bees.

*Each pound per acre is equal to approximately 1.1 kilograms per hectare.

PLANT AND NATIVE REGIONS	PLANT FAMILY, BLOOM, FRUIT AND COMPANIONS	HARDINESS	MAXIMUM HEIGHT × SPREAD AND GARDEN NOTES	BENEFITS TO BEES*
Chokeberry, Black *Aronia melanocarpa* **NW:** ON, QC, Great Lakes, Maritimes, northeastern US to Appalachian Mountains	E Rose. Clusters of dainty white shallow flowers. Bitter edible berries high in vitamin C; harvest after 2 frosts. Can spread aggressively. **Group with:** Elderberry, saskatoon berry.	**Perennial** Hardy to zone 3.	**6.5 ft (2 m) × 10 ft (3 m)** Sun or partial shade in poor to average, well-drained, wet to dry soils. Prune up to a third of old stems after flowering. Weeds and grasses inhibit growth.	**BB, HB, SB, BI** Visited by many bees for nectar and pollen including mason, mining, plasterer, sweat bees. Butterfly plant.
Chokecherry *Prunus virginiana* **NW:** Canada except YT, NU; northern US	E Rose. Racemes of small, white, shallow flowers. Almond-scented. Berries used to make wine.	**Perennial** Hardy to zone 1. Deer resistant.	**30 ft (9 m) × 20 ft (6 m)** Sun to light shade in a variety of well-drained soils, dry to moderate conditions. Prune in winter. Pioneer species.	**BB, HB, SB, BI** Single varietal honey is rare. Honey potential 22–45 lb/ac. Mining, sweat bees.
Cinquefoil, Shrubby, a.k.a. Potentilla *Dasiphora fruticosa* or *Potentilla fruticosa* **OW/NW:** Canada, northwestern US	M–L Rose. Shallow, bowl-shaped yellow 5-petal flowers. Cultivars have pink, yellow or red blooms. Foliage is silver green. **Group with:** Native roses, snowberry.	**Perennial** Hardy to zone 2. Deer and drought resistant.	**40 in (1 m) × 5 ft (1.5 m)** Sun to part shade, in poor to average, well-drained soil in dry to normal conditions. Prune after blooming. Divide every three years. Suitable for pond edge as well as hedgerow.	**BB, HB, SB, BI** Masked, sweat bees and more. Easy access, long-blooming plant suitable for the bones of a bee garden. Attracts many species of beneficial insects.
Cotoneaster, Rock *Cotoneaster horizontalis* **OW**	E Rose. Shallow, creamy white and pink blossoms. Red (inedible) berries and fall foliage. **Group with:** Shrubby cinquefoil, roses.	**Perennial** Hardy to zone 3. Deer and drought resistant.	**3 ft (0.9 m) × 5 ft (1.5 m)** Sun to light shade in a variety of well-drained soils. Prune in late summer or early fall. Can be trimmed to classic hedge shapes. To extend bloom period, plant several varieties.	**BB, HB, SB** Cuckoo, mining, sweat bees and more. Significant bumblebee shrub.
Cranberry, American *Vaccinium macrocarpon* **NW:** ON, QC, Maritimes	E Heather. "Shooting star" flowers with pink petals, black stamens. Nutritious red berries. **Group with:** Boneset to up bee value.	**Perennial** Hardy to zone 2.	**6 in (15 cm) × 1 ft (30 cm)** Sun to dappled shade in acidic, boggy soil. Pruning unnecessary. Suitable for pots with moist, peaty soil.	**BB, HB, SB** Light amber honey. Pollinated by 30+ species of native bees. Benefits with buzz pollination.
Cranberry, Highbush, a.k.a. Mooseberry *Viburnum edule* **NW:** Canada, northern US	E Adoxa. Clusters of shallow, white flowers. A few species have edible berries. **Group with:** Black chokeberry.	**Perennial** Hardy to zone 3.	**8 ft (2.4 m) × 6.5 ft (2 m)** Sunlight to part shade in moist, well-drained soil. Prune in late winter. There are many evergreen varieties bees love.	**BB, HB, SB** Mining and sweat bees. Many cultivars loved by bees.
Currant, Black *Ribes nigrum* **OW**	E Currant. Shallow pink and yellow flowers. Nutritious dark berries. **Group with:** Nodding onion, violets. Avoid white pine.	**Perennial** Hardy to zone 2. Deer resistant.	**5 ft (1.5 m) × 5 ft (1.5 m)** Sun to light shade in moist, well-drained soil. Need 2 varieties for pollination. Prune late fall to remove 4-year shoots. Water well.	**BB, HB, SB** Buzz-pollinated by bumblebees, especially queens.
Currant, Clove *Ribes aureum* **NW:** Canada, US	E Currant. Clusters of yellow flowers. Edible red berries.	**Perennial** Hardy to zone 2.	**10 ft (3 m) × 10 ft (3 m)** See black currant.	**BB, HB, SB** Long-tongued bees and hummingbirds.
Currant, Red-Flowering *Ribes sanguineum* **NW:** Pacific coast	E Currant. Pink blossoms. Edible red berries. **Group with:** Black currant.	**Perennial** Hardy to zone 6. Deer resistant.	**10 ft (3 m) × 5 ft (1.5 m)** Sun to part shade in rich, neutral to alkaline soil. Lightly prune after bloom. Water sparingly.	**BB, SB** Bumblebees love it; long-tongued bees.

PLANT AND NATIVE REGIONS	PLANT FAMILY, BLOOM, FRUIT AND COMPANIONS	HARDINESS	MAXIMUM HEIGHT × SPREAD AND GARDEN NOTES	BENEFITS TO BEES*
Dogwood, Red Osier *Cornus sericea* **NW:** Canada; all but southern US	E–M Dogwood. Clusters of shallow white flowers, protruding stamens. Red bark. **Group with:** Salal.	Perennial Hardy to zone 2.	13 ft (4 m) × 13 ft (4 m) Sun, variety of soils in normal to wet conditions. Prune in early spring. Suitable for pond edge as well as hedgerow.	BB, HB, SB, BI Long and short-tongued bees. Specialist pollinator: *Andrena fragilis*.
Elderberry, Black *Sambucus nigra* subsp. *canadensis* **NW:** MB, ON, QC, NS, NB, PEI, eastern US	E–M Adoxa. Large clusters of dainty white edible blossoms. Black berries edible fully ripe *and* cooked.	Perennial Hardy to zone 3. Deer resistant.	10 ft (3 m) × 10 ft (3 m) Sun or partial shade, fertile loamy soil, moist but well drained. Prune hard every third winter. Suitable for pond edge. Small carpenter, mason bees nest in hollow stems.	BB, HB, SB, BI Honey potential: 20–60 lbs/ac. Pollen: 600–800 lbs/ac. Small carpenter, sweat bees. Key pollen source for bees.
Goji Berry *Lycium barbarum* **OW**	M Nightshade. Light-purple blossoms. Flowers in year 2, edible red berries peak at 4–5 years.	Perennial Hardy to zone 5.	8 ft (2.4 m) × 13 ft (4 m) Sun. Well-drained moist soil rich in organic matter. Winter-tender until established. Prune lightly early spring. Avoid amendments. Needs air flow.	BB, HB Buzz-pollinated by bumblebees. Plant nodding onion around it for a bee-friendly combo.
Gooseberry, American *Ribes hirtellum* **NW:** Canada except BC; northeastern US	E Currant. Small cup-shaped flowers with greenish-white blossoms. Edible berries. Avoid pine.	Perennial Hardy to zone 3.	40 in (1 m) × 40 in (1 m) Sun to part shade, amended soil, good drainage, water in dry periods. Protect from heat. See European gooseberry.	BB, SB Mining and sweat bees. Special value to native bees. See European gooseberry.
Gooseberry, European *Ribes uva-crispa* **OW**	E Currant. Pink or white shooting-star flowers, long stamens; more nectar than currant. Edible berries. Avoid pine.	Perennial Hardy to zone 3.	5 ft (1.5 m) × 5 ft (1.5 m) Prune current season's growth 3 buds from base in late fall. See American gooseberry.	BB, HB, SB Honey plant; however, honey-bees prefer cherry blossoms. Yellow-green pollen. See American gooseberry.
Huckleberry, Evergreen *Vaccinium ovatum* **NW:** BC coast	E–M Heather. Clusters of 3–10 pink bell-shaped blossoms. Harvest edible blue-purple berries after first frost. **Group with:** Blueberry.	Evergreen Perennial Hardy to zone 7. Deer resistant.	8 ft (2.4 m) × 10 ft (3 m) Sun to dappled shade. Acidic soil, rich in humus. Once plant is established, water deeply but less frequently. Prune to remove damage.	BB, SB Bumblebees and other long-tongued bees.
Kinnikinnick, a.k.a. Bearberry *Arctostaphylos uva-ursi* **OW/NW:** Canada, northern and western US	E Heather. Clusters of white to pink urn-shaped flowers. Edible red berries. **Group with:** Plant as understory for shrubs and trees.	Perennial Hardy to zone 2.	1 ft (30 cm) × 13 ft (4 m) Sun to shade, in rocky or sandy, acidic soils. Do not fertilize. Prune spring or summer. Propagate cuttings. Mulch. Do not disturb sensitive roots.	BB, HB, SB Bumblebees (especially queens), digger, mason and mining bees. Mostly nectar.
Lingonberry, a.k.a. Cowberry *Vaccinium vitis-idaea* subsp. *minus* **OW/NW:** Canada	E Heather. Pink, bell-shaped flowers. Edible red berries. **Group with:** Plant as understory.	Perennial Hardy to zone 3 with winter protection.	1 ft (30 cm) × 20 in (50 cm) Cannot tolerate alkaline soils. Needs good drainage, frost protection, prefers some shade. Hardy; rarely needs pruning.	BB, HB, SB Mining bees love the nectar.
Mock Orange *Philadelphus* spp. **OW/NW:** BC, AB, US Pacific coast, ID, MT	E–M Hydrangea. Fragrant, shallow, white flowers with yellow stamens. **Group with:** Bleeding heart, masterwort.	Perennial Hardy to zone 3. Deer resistant.	5 ft (1.5 m) × 5 ft (1.5 m) Sun to dappled shade in well-drained soil. Prune for stronger stems, more blossoms. Hedgerow or grow as feature.	BB, HB, SB, BI Important for filling June forage gap for bees. Avoid double-blossom varieties.

PLANT AND NATIVE REGIONS	PLANT FAMILY, BLOOM, FRUIT AND COMPANIONS	HARDINESS	MAXIMUM HEIGHT × SPREAD AND GARDEN NOTES	BENEFITS TO BEES*
Ninebark, Eastern *Physocarpus opulifolius* **NW:** ON, QC, MB, NS, NS, eastern US	M Rose. Globular clusters of shallow white blossoms. 'Diablo'—burgundy foliage. **Group with:** Serviceberry.	Perennial Hardy to zone 2.	**10 ft (3 m) × 5 ft (1.5 m)** Sun to partial shade, well-drained soil. Adaptable. Prune June after flowering if needed.	**BB, HB, SB, BI** Bee garden essential. Easy access blooms. Special value for native bees. Butterfly plant.
Ninebark, Mallow *Physocarpus malvaceus* **NW:** BC, AB	M Rose. See eastern ninebark. **Group with:** Native rose, saskatoon berry, spiraea.	Perennial Hardy to zone 2.	**10 ft (3 m) × 10 ft (3 m)** See eastern ninebark.	**BB, HB, SB, BI** See eastern ninebark. Mining bees and more. Good blossom density with high nectar and pollen rewards.
Oceanspray *Holodiscus discolor* **NW:** BC, northwest US	M Rose. Tiny, fragrant, creamy white flowers. **Group with:** Nootka rose, snowberry.	Perennial Hardy to zone 5. Drought resistant.	**13 ft (4 m) × 13 ft (4 m)** Sun, partial shade in gravelly or rocky medium to moist soil. Prune back hard after flowering to desired size/shape.	**BB, HB, SB, BI** Accessible blossoms help to fill the June forage gap for all bees. Butterfly host plant.
Oregon Grape, Dull *Mahonia nervosa* **NW:** Pacific Coast	E Barberry. Clustered yellow flowers, shallow cups. Edible purple berries. **Group with:** Plant as understory.	Perennial Hardy to zone 5.	**1 ft (30 cm) × 4 ft (1.2 m)** Prefers partial or light shade, but tolerates sun if the soil doesn't dry out. Well-drained soil rich in humus .	**BB, HB, SB** Mason bees. Significant bumblebee plant for nectar and pollen.
Oregon Grape, Tall *Mahonia aquifolium* **NW:** BC and AB to CA; ON, QC, Great Lakes	E Barberry. See dull Oregon grape. The clumps of edible purple berries sweeten after frost.	Perennial Hardy to zone 5.	**6.5 ft (2 m) × 5 ft (1.5 m)** See dull Oregon grape. Prune to desired height in spring after flowering.	**BB, HB, SB** See dull Oregon grape. Essential early-blooming shrub for blue orchard mason bees.
Pieris, Japanese, a.k.a. Lily of the Valley Shrub *Pieris japonica* **OW**	E Heather. Clusters of small white to pink bell-shaped flowers with long corollas. **Group with:** Oregon grape as hedgerow under-shrubs.	Evergreen Perennial Hardy to zone 6.	**10 ft (3 m) × 6.5 ft (2 m)** Part shade in moist, rich, acidic soil. Prune damage in spring. Deadhead to encourage blooms. Needs winter protection in colder zones.	**BB, SB** Blooming indicates time to release mason-bee cocoons. Mining bees.
Plum, Indian *Oemleria cerasiformis* **NW:** Pacific Coast	E Rose. Drooping clusters of creamy white, shallow blossoms. Small edible purple fruits.	Perennial Hardy to zone 7. Drought resistant.	**16 ft (4.9 m) × 13 ft (4 m)** Sun or light shade in moist, rich soil. Best left unpruned, but remove suckers. Plant the seeds to grow new bushes.	**BB, HB, SB** Male and female flowers on separate bushes provide key early-spring forage.
Raspberry, Black *Rubus occidentalis* **NW:** ON, QC, NB; US north of TX and FL	E Rose. Shallow blossoms with multiple stamens and tiny white petals. Delicious and nutritious red berries. **Group with:** Blackberry, native rose.	Perennial Hardy to zone 4.	**10 ft (3m) × 3 ft (0.9 m)** Purchase bare-root canes in late winter or early spring to plant while dormant. Prune out older canes once a year. May need support.	**BB, HB, SB, BI** Honey potential 45–90 lbs/ ac. Cuckoo, leafcutter, mason, mining, sweat bees. Whitegrey pollen with good protein content.
Raspberry, Wild Red *Rubus idaeus* **NW:** Canada; continental US except southeast	E Rose. See black raspberry.	Perennial Hardy to zone 3.	**5 ft (1.5 m) × 5 ft (1.5 m)** See black raspberry.	**BB, HB, SB, BI** See black raspberry.
Rose, Arctic, a.k.a. Prickly Rose *Rosa acicularis* **NW:** Canada except NL and NU; north, central US	M Rose. Alberta's floral emblem. See rugosa rose.	Perennial Hardy to zone 2.	**10 ft (3 m) × indefinite** See rugosa rose.	**BB, HB, SB, BI** See rugosa rose.

PLANT AND NATIVE REGIONS	PLANT FAMILY, BLOOM, FRUIT AND COMPANIONS	HARDINESS	MAXIMUM HEIGHT × SPREAD AND GARDEN NOTES	BENEFITS TO BEES*
Rose, Nootka *Rosa nutkana* **NW:** Pacific coast, western US from AK to CA	**M** Rose. Pink flowers sometimes buzz-pollinated. Edible petals and hips. **Group with:** Shrubby cinquefoil.	**Perennial** Hardy to zone 3.	**8 ft (2.4 m) × 5 ft (1.5 m)** Apply aged compost and manure spring and fall. Prune only as necessary in spring. See rugosa rose.	**BB, HB, SB, BI** Loved by bumblebees. Plant varieties with varying bloom times. See rugosa rose.
Rose, Rugosa *Rosa rugosa* **OW**	**M** Rose. Shallow-cupped blooms, numerous stamens. Petals white, yellow, pink or red. Often fragrant. Edible petals and hips.	**Perennial** Hardy to zone 3. Drought resistant.	**5 ft (1.5 m) × 5 ft (1.5 m)** Sun in rich, moist, well-drained soil, but tolerates poor soil. Plant in spring. Needs good air circulation. Avoid wetting leaves. Avoid double blossoms. See Nootka rose.	**BB, HB, SB, BI** Honey potential: 20–30 lbs/ac. Pollen: 40–60 lbs/ac. Roses are important to help fill the June nectar and pollen gap for many bees. See Nootka rose.
Salal *Gaultheria shallon* **NW:** Pacific coast	**M** Heather. Pink to white urn-shaped flowers. Edible blue-black fruit. **Group with:** Evergreen huckleberry, wintergreen.	**Perennial** Hardy to zone 6. Deer and drought resistant.	**5 ft (1.5 m) × 5 ft (1.5 m)** Sun to deep shade in variety of soils. Moist or dry conditions. Prune in spring. Suitable for hedgerow understory, edible forest, ground cover.	**BB, HB** Primarily visited by bumblebees for nectar. Heritage honey plant in BC.
Salmonberry *Rubus spectabilis* **NW:** Pacific coast	**E** Rose. Shallow fuchsia flowers with 5 petals. Edible berries June–July. **Group with:** Blackberry and thimbleberry.	**Perennial** Hardy to zone 5.	**10 ft (3 m) × 10 ft (3 m)** Sun to partial shade in moist soil. Prune after fruiting and irrigate in dry periods. Beautiful in hedgerow, edible forest.	**BB, HB, SB** Flowers loved by bumblebee queens building up brood. Mason bees provision nests with the pollen.
Saskatoon Berry *Amelanchier alnifolia* **NW:** Canada except NU and Maritimes; AK, northwest US	**E** Rose. Clusters of white flowers with 5 petals. Purple berries midsummer for delicious pies. **Group with:** Underplant with crocus.	**Perennial** Hardy to zone 1. Deer and drought resistant.	**13 ft (4 m) × 13 ft (4 m)** Sun to partial shade in fertile, humus-rich, well-drained acidic soil. Prefers moist conditions. Prune young plants so only strongest stems remain.	**BB, HB, SB, BI** Mining and sweat bees. Mostly pollen. An early spring source of food for bees.
Serviceberry *Amelanchier canadensis* **NW:** ON, QC, NS, NB, PEI, US Atlantic coast	**E** Rose. Clusters of white flowers with 5 petals. **Group with:** Underplant with crocus.	**Perennial** Hardy to zone 3. Deer resistant.	**30 ft (9 m) × 30 ft (9 m)** See saskatoon berry.	**BB, HB, SB, BI** Special value to native bees.
Silver Buffaloberry *Shepherdia argentea* **NW:** Prairies, western US	**E–M** Oleaster. Inconspicuous shallow yellow blossoms. Male and female blossoms on separate shrubs. Berries edible in small amounts; sweeten after several frosts. 'Goldeye' cultivar developed in MB; 'Sakakawea' in US.	**Perennial** Hardy to zone 1.	**20 ft (6 m) × 13 ft (4 m)** Sun in moist, well drained, rocky soil. Prune suckers. Propagate by seed, cuttings, suckers. Wind tolerant. Fixes nitrogen, stabilizes soil. Suitable for xeriscape hedgerow, edible forest. First Nations food plant.	**BB, HB, SB** One of the first honey plants to bloom on the prairies. Important food for native bees and honeybee spring brood.
Snowberry, Common *Symphoricarpos albus* **NW:** BC to QC, NWT, AK to CA, northern US	**E–M–L** Honeysuckle. Small pink bell-shaped flowers. Toxic white berries. Berries are toxic. **Group with:** Native roses, shrubby cinquefoil.	**Perennial** Hardy to zone 3. Deer resistant.	**4 ft (1.2 m) × 5 ft (1.5 m)** Sun to partial shade in clay or sand. Dry to moist conditions. Remove suckers. Prune in winter. Allow air circulation to prevent powdery mildew.	**BB, HB, SB** Honey plant. Essential bee plant covers gaps in bloom spring to fall.
Soapberry *Shepherdia canadensis* **NW:** Canada; northwestern US	**E–M** Oleaster. See silver buffaloberry. **Group with:** Native roses, yarrow.	**Perennial** Hardy to zone 2.	**8 ft (2.4 m) × 8 ft (2.4 m)** See silver buffaloberry. Also called "Soopolallie"; a traditional First Nations food, the berries can be whipped into foamy "ice cream."	**BB, HB, SB** Honey plant. Important shrub for native bees.

PLANT AND NATIVE REGIONS	PLANT FAMILY, BLOOM, FRUIT AND COMPANIONS	HARDINESS	MAXIMUM HEIGHT × SPREAD AND GARDEN NOTES	BENEFITS TO BEES*
Spiraea, Douglas, a.k.a. Hardhack *Spiraea douglasii* **NW:** Pacific coast, ID, MT	M–L Rose. Fluffy spikes of pink blossoms. Spreads aggressively. **Group with:** Native roses, oceanspray.	Perennial Hardy to zone 5.	8 ft (2.4 m) × 10 ft (3 m) Sun to part shade in moist, well drained, acidic soil. Prune hard in spring. Deadhead. Can spread aggressively; cut back hard after blooming.	BB, HB, SB, BI Bumblebees scramble for pollen over the surface of this bee "cotton candy."
Spiraea, Japanese *Spiraea japonica* 'Crispa' **OW**	M Rose. Masses of white to pink or red blossoms (bees favour the white). Avoid double blossoms. Choose non-invasive cultivars. **Group with:** Scarlet bee balm.	Perennial Hardy to zone 3.	3 ft (0.9 m) × 3 ft (0.9 m) Sun to partial shade and moist, rich soil, but can tolerate a range of well-drained soil. Prune in winter.	BB, HB, SB, BI Bumblebees scramble over blooms collecting pollen. Helps fill June forage gap for native bees.
Sumac, Smooth *Rhus glabra* **NW:** BC to QC, continental US	E–M Cashew. Brilliant clusters of red velvety flowers. Female and male flowers on separate shrubs. Edible red berries used to make lemonade. Spreads aggressively. **Group with:** Pin cherry and chokecherry.	Perennial Hardy to zone 3. Deer and drought resistant.	13 ft (4 m) × 13 ft (4 m) Full to partial sun in well-drained soil. Prune in early spring only if necessary. Suitable for prairie hedgerow. Pioneer species with great fall colour.	HB, SB, BI Honey plant. Male shrubs offer nectar and pollen, female shrubs nectar. Mining, plasterer, small carpenter, sweat bees. Nesting material.
Tea, Labrador *Ledum groenlandicum* **OW/NW:** Canada and northern US	E Heather. Clusters of shallow white flowers, protruding stamens. Toxic plant; tea must be weak and limited. **Group with:** Salal.	Perennial Hardy to zone 1. Deer resistant.	4 ft (1.2 m) × 3 ft (0.9 m) Sun to shade in moist to wet, peaty, acidic soil. Suitable for pond edge. Small doses of weak tea from leaves used to treat colds and stomach ailments.	BB, HB, SB, BI Honey plant. Important to bumblebees and the mining bees that pollinate blueberries in the Maritimes.
Tea, New Jersey *Ceanothus americanus* **NW:** ON, QC, eastern US	E–M Buckthorn. Clusters of white flowers. Leaves used for tea during American Revolution. **Group with:** Blazing star.	Perennial Hardy to zone 4. Drought resistant.	4 ft (1.2 m) × 5 ft (1.5 m) Sun or light shade in moist, sandy or loamy soil. Fixes nitrogen. Prune to shape. Dye plant. Essential bee plant.	BB, HB, SB, BI Honey plant. Mining, plasterer, sweat bees. Hummingbird and butterfly shrub.
Thimbleberry *Rubus parviflorus* **NW:** BC, AB, ON, Great Lakes, western US, Pacific coast	E Rose. Shallow flowers with white petals. Edible pink-red berries high in vitamin C. **Group with:** Ninebark.	Perennial Hardy to zone 3.	8 ft (2.4 m) × 8 ft (2.4 m) Sun to partial shade in clay, sand and loam. Keep moist. Prune fruiting canes after finished producing each year.	BB, HB, SB Honey plant. Important easy-access forest-edge plant for bees, notably bumblebees.
Wintergreen, a.k.a. Partridgeberry, Eastern Teaberry *Gaultheria procumbens* **NW:** MB and eastern Canada, eastern US	E Heather. Pink and white bell-shaped blossoms. Edible red berries best after a frost. **Group with:** Heather, salal.	Perennial Hardy to zone 3. Deer resistant.	8 ft (2.4 m) × 3 ft (0.9 m) Light or dappled shade. Will tolerate sun if soil kept moist. Soil neutral to acidic with humus. Prune after flowering. Spreads by rhizomes. Seed requires moist stratification.	BB, HB, SB Mainly visited by bumblebees seeking nectar. Beekeepers are experimenting with wintergreen oil to eradicate varroa mites.

*Each pound per acre is equal to approximately 1.1 kilograms per hectare.

Victory Borders for Bees

CHAPTER NINE

*I*n a Victory Border for Bees, there is no room for slackers or sissies. We want drought-tolerant, long-blooming, tough-love plants with staying power. We want abundant nectar and pollen, and we need bee plants with long-term commitment. There are no "bloom today and gone tomorrow" plants in the Victory Borders for Bees. The less maintenance these plants demand, the less mucking around and fussing the gardener needs to do, which means less risk of tramping on ground-dwelling bees and their nest entrances. And the bigger the border, the better. Bees need room to feed and breed, and a well-planted Victory Border gives bee boys and girls a place to dine, dance, date, mate and raise new troops of pollinators.

FOOD SECURITY FOR BEES

With the trend toward growing annuals for food, it's more important than ever to create a framework of perennials around our food gardens to provide a steady flow of nectar and pollen for bees. This is where those old dependables come in, the long-blooming perennials. These pollen-packing, nectar-rich plants provide emergency rations in times when there is a dearth of blooms from trees, shrubs, vegetable patches and other plants, filling in gaps and helping developing bees stay strong and healthy.

Opposite: Globe thistle is an excellent drought-tolerant perennial for bees, especially grouped with catmint, Russian sage, sea holly and lavender.

Planting a large swath of lavender, like this border at Vancouver's City Hall, makes excellent bee pasture. Different varieties of lavender bloom in succession, attracting different guilds of bees.

PLANTING ☀ PLAN ❀

A Boulevard Border for Bees: Drought-tolerant, Low-maintenance Bee-scaping

Replacing lawns with drought-tolerant flowers will beautify your block and save money on water and maintenance. A collective effort on replacing lawns with bee borders will have a dramatic impact on biodiversity in your neighbourhood. Boulevard bee gardens give bees and butterflies traffic zones of their own, with paths of painterly perennials. They give cyclists and pedestrians beautiful dynamic routes that transform with the seasons. Creating and installing an art project with garden-themed sculptures will help buffer the pollution on the edge closest to passing cars and further boost the benefits for bees. See opposite page for a sample boulevard border gardening plan.

When choosing perennials, take the no-frills approach and look for hardy, low-maintenance heritage cultivars or native plants that have a long history of attracting bees. Avoid the new gaudy double or frilly cultivars that are of trifling use to pollinators, often with little or no nectar or pollen rewards. Many of these plants were created from wildflowers that had a genetic anomaly, where the pollen-producing parts had mutated into petals. Plant selection and breeding can change the morphology, colour and pollen and/or nectar rewards that the plant evolved over millennia to attract pollinators. In some cases, cultivars do offer more rewards for bees because they have an extended bloom time and/or more flowers than the wild species. A good example of this is shrubby penstemon 'Purple Haze' (*Penstemon fruticosus*), which has a greater profusion of purple blooms than the wild plant. And in some cases, heritage cultivars are more appropriate for the home garden because the wild variations (especially pioneer species) are aggressive and begin to crowd out other plants. Such is the case with fireweed (*Chamerion angustifolium*), which can take over a garden; the white-flowered variety called 'Album' is a non-aggressive cultivar that still brings in the bees.

MAKE YOUR BEE BORDER DESIGN BUZZ WITH POSSIBILITY

When you are designing your Victory Border for Bees, choose short, medium and tall flowers for small, medium and large bees. You are setting up a canteen

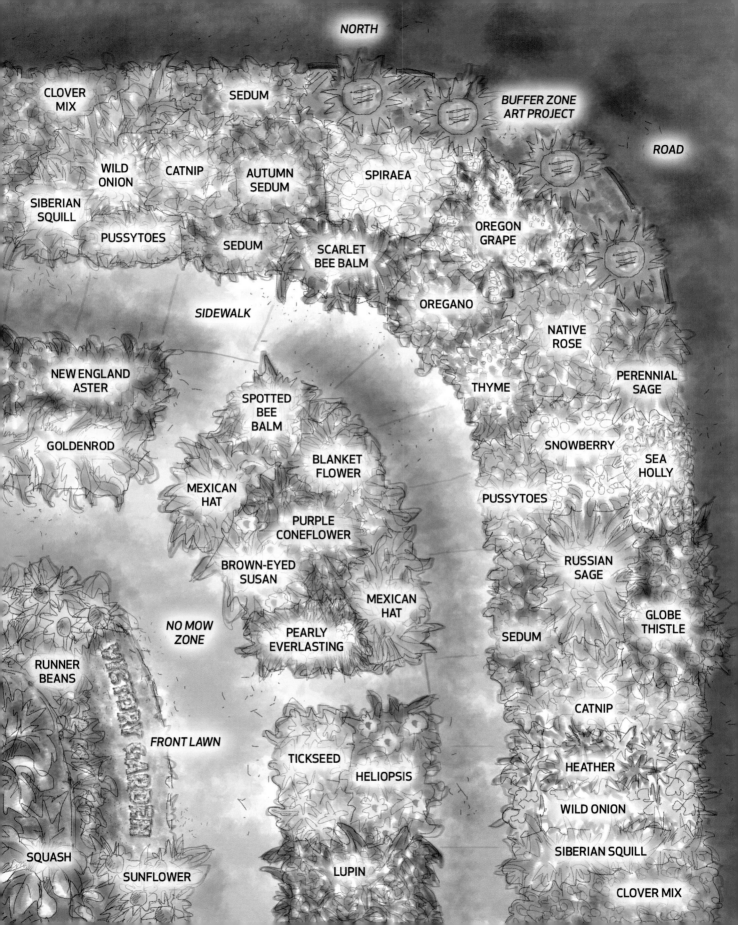

for bees of all stripes—from those tiny, short-tongued halictid bees to medium-sized mason bees and majestic, long-tongued bumblebee queens. Here are tried-and-true tips to make your bee border buzz with life:

- Avoid "one-itis"—stand-alone plants—in favour of clusters of three, five, seven and so on, which provide a beautiful presentation and are better for bees. Case in point: growing a single lupin in my garden attracted a few bumblebees; planting three attracted several species of bees.

- Think of your garden as a quilt and scale a patchwork of flower drifts accordingly. An ideal "patch" is 3 square feet (0.9 square metres). Or if you have a small garden, try dwarf or alpine varieties in tiny clumps. Large patches and drifts of flowers create a critical mass of blooms that are alluring to bees and require less energy from them to "switch gears" as they move from one flower type to another. And ground-nesting bees are more likely to create their brood chambers near undisturbed, dependable and bountiful sources of food.

Whether you are seeking a calm respite from your busy life or a coral reef drenched with colour and sensuous texture, large drifts of pollinator-sustaining perennials are the bones of a successful Victory Garden for Bees. Plant in sumptuous clumps and ensure that each garden has a triple dose of bee-friendly flowers to support bees of all types.

A PEACE GARDEN FOR BEES: WHITE POPPIES, MILKWEED AND MISS WILLMOTT'S GHOST

A pollinator garden is a sacred place that honours creatures inextricably connected to our collective fate. A soothing collection of white blossoms invokes a

A honeybee forages on 'Miss Willmott's Ghost' sea holly. This particular flower sports an odd genetic mutation that caused the two halves to present different colors.

calmness of spirit and meditation while also providing heavenly texture and scent. This peaceful offering to pollinators is also the perfect place for people—set up a shady bench to rest from a world often overwhelming with colour and chaos. Although many bee plants are yellow, blue and purple, there are numerous white flowers that bees see as "bee violet." The colour ultraviolet, which bees can see but humans cannot, is sometimes called "bee violet," and when this colour is mixed with a greenish-yellow hue it is called "bee purple." This means bees see white flowers as bee violet and yellow flowers as bee purple. White oriental poppies (*Papaver orientale* 'Royal Wedding'; see page 86 for planting advice) are among my favourite flowers, often providing the striking sight of a bee covered in velvety black pollen and hovering about the dramatic black and white blooms. Varieties of white-flowering milkweed, such as *Asclepias incarnata* 'Ice Ballet' (see page 128 for planting advice), invite butterflies and bees to sip together in your oasis. Another favourite

of mine, created by a master of the white garden, is celebrated garden-designer Gertrude Jekyll's sea holly cultivar 'Miss Willmott's Ghost' (*Eryngium giganteum* 'Miss Willmott's Ghost'; see blue sea holly on page 201 for planting advice). For another tranquil triad of bumblebee blossoms, combine white lupin (*Lupinus albus*), white obedient plant (*Physostegia virginiana* 'Crystal Peak White') and white false indigo (*Baptisia lactea*; see false blue indigo on page 127 for planting advice).

COTTAGE-GARDEN TRIO: MALLOW, MASTERWORT AND PERENNIAL SAGE

Like their taller cousins the hollyhocks, zebra mallows are classic cottage-garden plants. The pink or violet flowers with darker "zebra stripes" are an excellent illustration of nectar guides that humans can clearly see. *Malva sylvestris* 'Zebrina' is a heritage striped mallow that comes true from seed and supports bees from June to September, making it a gorgeous gap-filler. Thriving on neglect, it's a good bet for that part of the garden where nothing else blooms. In fact, *sylvestris* is Latin for "woods" and zebra mallow is great as an understory in a hedgerow as well as in the Victory Border; give it a little legroom because it is a heavy feeder, stealing nutrients and moisture from neighbours. This is a trifecta plant, benefitting bees, hummingbirds and butterflies. Up the trio effect by adding two more of the best long-blooming perennials for bees: masterwort (*Astrantia major*) and lilac sage (*Salvia verticillata*).

CORAL REEFS FOR BEES: ASTER, SEDUM AND FLEECEFLOWER

UK gardener Marc Carlton paints a picture of his biodiverse collection of perennials: "My pollinator border

is a 'terrestrial coral reef,' not only a mass of disparate shapes and colours, but alive with the movement of small creatures—but insects, not fish."[56] Three late-season plants that form a vibrant coral reef of colour and texture are sedums, asters and fleeceflowers. Even as the leaves turn gold and red and bumblebee queens prepare to hibernate, these plants shimmer brilliantly in your garden and give a fall boost to honeybees and bumblebees before they hunker down for winter. Bumblebees love late-summer sedums so much that you will see them frenetically fumbling with the buds even before they open. *Sedum telephium* 'Purple Emperor' is a cultivar with lovely dark-burgundy foliage. A heritage cultivar with a strongly documented pull for bees is 'Mönch' aster (*Aster × frikartii* 'Mönch'). This robust cultivar is popular with a host of insects, including many of the small native bees. 'Firetail' mountain fleeceflower (*Persicaria amplexicaulis* 'Firetail') attracts bees in late fall when there is not much else blooming; for another wave of colour, add drumstick

Container Pit Stops for Bees

Provide groups of prolific bloomers to fuel bees as they forage so they don't get stranded, running short of energy on the way back and forth from their nest. Generous pots of perennials can revive and revitalize travel-weary pollinators.

A Peace Garden in Pots

⚜ White lavender (*Lavandula angustifolia* 'Nana alba')
⚜ Garlic chives (*Allium tuberosum*)
⚜ Greek oregano (*Origanum vulgare*)

Cottage Perennials in Pots

⚜ *Campanula carpatica* 'Deep Blue Clips'
⚜ *Campanula punctata* 'Summertime Blues'
⚜ *Campanula medium* 'Deep Blue'

Coral Reef Perennials in Pots

⚜ Red snapdragon (*Antirrhinum majus* 'Black Prince')
⚜ Burgundy masterwort (*Astrantia major* 'Ruby Wedding')
⚜ 'Burgundy' blanketflower (*Gaillardia × grandiflora* 'Burgundy')

Above: The white poppy has become an emblem of peace and remembrance of all victims of war. **Opposite, clockwise from top left:** On a November day at Vancouver's VanDusen Botanical Garden, worker bees head straight for the bistort. Hollyhock is a pollen-rich biennial much loved by bumblebees, which sleep in the flowers at night. The pollen grains of the mallow species are very large, at 144 micrometres in diameter. Plant a variety of bellflowers with different sizes and bloom times to attract different sizes and species of bees.

allium (*Allium sphaerocephalon*) with its intensely purple pollen and blossoms.

CLEAN AS YOU GROW, GOING WITH THE BEE FLOW

There's a time to plant and a time to weed. There's a time to trim and a time to compost. There's a time to prune and a time to divide perennials. And there's a time when it is best to just to stay away from the garden and leave it be for the sake of the bees. If we listen to the bees and plan our interventions around their schedules, our gardens will be refuges for them; if we trim hollow stems where bees are hibernating, then we won't be helping them. Over time, you will develop an awareness of the life cycles of the bees to help guide your gardening interventions. Think about how native bees live in the wild, nesting in the dry stems from the previous winter even though they will break down naturally over time. Go with the bee flow and your Victory Border will prosper along with the bees.

BOOST YOUR GARDEN'S FLOWER POWER

Here are a few tricks to up the nectar and pollen output of your perennial plantings for bees.

∽ Most flowers do benefit from deadheading so they

Opposite, clockwise from top left: *Verbena bonariensis* can be integrated into a coral reef for bees and extend bloom succession into the fall. Late-blooming sedum is a favourite plant for honeybees and bumblebees. Plant Siberian squill in the fall to provide spring blooms with blue pollen for early bees like this little sweat bee (*Lasioglossum*). If you like dahlias, plant single varieties that provide some rewards for bees, such as the striking 'Bishop of Llandaff'. **Above:** Sea holly is one of the best perennials for bees.

can do a repeat bloom, a "clean as you grow" task.

∽ Shrubby perennials, like germander, benefit from pruning after blooming because they will branch out and bloom again later in the same season, even more profusely. Remember to never cut back a perennial plant more than a third at a time.

∽ In a large patch of perennials, cutting a third back from one section before flowering will delay that section and help to stagger bloom times. This will help prevent a bloom gap, especially in years when blooming has been delayed by a cold spring and then everything seems to bloom at once.

The Laburnum Walk at VanDusen Botanical Garden buzzes with bumblebees foraging in the yellow laburnum blossoms and the purple globes of ornamental alliums.

~ When the plant is blooming, make sure it is getting enough water to produce abundant nectar.

BULBS FOR BEES

In the fall, plant a fairy ring of bulbs in your lawn or around a tree in the boulevard, starting with spring crocuses (*Crocus vernus*) and Siberian squill (*Scilla siberica*). Naturalizing crocuses will go forth and multiply, so you can spread the love, sharing bulbs with friends and neighbours. Perform some interventions within the drip lines of trees—make these into no-mow zones and load them up with bulbs, including the dramatic blooms of blue camas (*Camassia quamash*; planting information on page 123). Edible and ornamental onions will take it from there and bloom in succession in mid- and late summer. If you want to make sure squirrels don't dig up your bulbs, lay some chicken wire down over the planting until they start poking through in the spring.

Be absolutely sure you are not buying bulbs that have been treated with neonicotinoids, which make the flowers toxic for bees.

CUT, ROOT AND SHARE: HOW TO PROPAGATE PERENNIALS FROM CUTTINGS

Once you've established a good bee-feeding perennial garden, grow more of the same and share the bounty with friends by propagating cuttings. Shrubs can be rooted directly into outdoor beds with good drainage and then transplanted to a permanent home once they have successfully rooted. (Choose an area with well-drained soil, dig in some grit and proceed as below.) Once it has taken root, carefully dig out the plant and replant in its new location. Rooting pots of perennials on windowsills is a great indoor winter garden project. Bee plants that root well from cuttings include asters, lavender, mint, sage, sedum and speedwell.

1. Start with a sterile growing medium with good drainage. I suggest one part coarse sand to one part vermiculite.

Opposite, clockwise from top left: Bumblebees and large leafcutter bees are well supported by perennial sweet peas (*Lathyrus latifolius*). *Scabiosa stellata* 'Ping Pong' has demure blue-violet flowers that fill the garden with a marshmallow-honey fragrance that bees love. Zinnias are like fireworks for the later summer and fall garden. Long-blooming pincushion flowers are available as annuals or perennials in many colours.

2. Make a rooting hormone tea by pouring boiling water over willow twigs and letting it steep for a day, or buy it from the store.

3. Choose a plant that is healthy and leafed out, but not blooming. Snip off a stem 6 inches (15 cm) from the tip, cutting just below a leaf node.

4. Snip off the leaves on the bottom 2 inches (5 cm) of the stem.

5. With a chopstick, make a hole in the growing medium 2 inches (5 cm) deep.

6. Put a bit of rooting hormone tea in a small container and dip the bottom part of the stem in it. Insert the stem in the growing medium to just below the bottom leaf and fill in the soil.

7. Tie a plastic bag over the plant to preserve moisture, keeping the medium moist but not sopping wet.

8. Look at the bottom of the plant after about three weeks to see if it is rooting, or gently pull up to test if it has rooted. Once the first new leaves appear, it is ready to be hardened off and transplanted.

LONG-BLOOMING ANNUALS THAT LOVE PERENNIALS

Even if you have a continuously blooming border of perennials for bees, there are some ornamental annuals that help boost the amount of nectar and pollen during summer and early fall, which is when the population of bees is largest.

The plush centre of elecampane provides a luxurious landing and refuelling pad for pollinators like this long-horned bee (*Melissodes*).

The filaments are the shafts of the pins, and the anthers the pin heads upon which the pollen sits daintily until an insect lands and scatters it like dust. Pincushion flowers have a long bloom time (8 to 13 weeks) in late summer, making them an important succession plant. The nectaries vary from shallow to deeper and are a perfect fit for short and medium-tongued native bees in the New World.

MEXICAN SUNFLOWER: LATE-SUMMER FIREWORKS

The bright orange-red blossoms of Mexican sunflower (*Tithonia* spp.) go very nicely next to scarlet runner beans and the fiery blossoms of scarlet bee balm. In late summer and early fall, Mexican sunflower is bustling with leafcutter bees, bumblebees and hummingbirds. Zinnia and dahlia make two great companions for it, particularly velvety scarlet zinnias and 'Bishop of Llandaff' dahlias, with petals that start out scarlet and fade to coral against dark bronze foliage.

ZINNIA: BEE FLOWERS FOR DRAMATIC COLOUR

Zinnias (*Zinnia elegans*) are very showy late-summer flowers. Again, choose single heritage varieties to provide optimum succour for bees. The larger flowers are more attractive than dwarf variations, and hummingbirds definitely prefer red. Whatever colour you pick, bumblebees will love to nestle down in the flowers for a quiet nap until they wake up again, sip some nectar and get back to work.

HONEYWORT: BUZZ-POLLINATED BELL BLOSSOMS

One of the rare plants I grow in my garden that is not edible and medicinal, honeywort (*Cerinthe major* 'Purpurascens') is in the borage family and is also known as "blue shrimp plant." I love the purple and teal foliage, and it gives me pleasure to hear the bumblebees buzz-pollinating inside the bell-shaped blossoms. Like borage, it will reseed itself, but is well-behaved. Imagine fields of these growing wild in southern Italy or Greece, with all their colour variations ringing with choruses of bees.

PINCUSHION FLOWER: VELVETY PISTIL PINCUSHIONS

Pincushion flower (*Scabiosa atropurpurea*) is so named because the male parts of the plant— pistils—look just like pins sticking out of a neatly domed flower head.

PERENNIALS
for Bees

A SEASONAL PLANTING CHART

$\mathcal{A}$ bee garden benefits from a mix of native, near-native (from surrounding regions) and exotic perennials. Indigenous bees prefer native plants but also benefit from long-blooming exotics that extend the bloom season of flowers and help fill any nectar and pollen gaps. Choose different varieties with staggered bloom times and avoid double blossoms. Plant zones can be extended with winter protection. Be aware that perennials in pots should be considered a zone or more tender than those in the ground.

PLANT ORIGIN
OW: Old World (indigenous to Europe, Asia or Africa)
NW: New World (indigenous to the Americas and their islands)

BLOOM PERIOD
E: Early
M: Mid-season
L: Late
SUC: May be seeded in succession

HARDINESS
refers to the coldest zone the plant is hardy to.
DEER AND DROUGHT RESISTANCE are most reliable once the plant is established.

BEES ATTRACTED
BB: Bumblebees
HB: Honeybees
SB: Solitary bees
BI: Beneficial insects

NAME AND NATIVE REGIONS	PLANT FAMILY, BLOOM, FOLIAGE AND COMPANIONS	HARDINESS	MAXIMUM HEIGHT AND GARDEN NOTES	BENEFITS TO BEES*
Allium, Blue Globe *Allium azureum* OW	E–M Amaryllis. Blue spheres. **Group with:** Blue camas, nodding onion, sage, woolly lamb's ear.	**Perennial** Hardy to zone 4. Deer resistant.	**2 ft (60 cm)** Sun to part shade. Well-drained, alkaline to neutral soil, moderate moisture. Plant in fall, 3 times as deep as bulb height.	**BB, HB, SB** Mainly nectar rewards. Butterfly plant. Support bees with varied allium cultivars for extended bloom.
Allium, Drumstick *Allium sphaerocephalon* OW	M Amaryllis. Egg-shaped heads of small, purple blossoms. See blue globe allium.	**Perennial** Hardy to zone 4. Deer resistant.	**2 ft (60 cm)** Blooms later than most other ornamental onions. Suitable for pots. See blue globe allium.	**BB, HB, SB** Purple pollen. See blue globe allium. Significant bumblebee plant.
Allium, 'Purple Sensation' *Allium aflatunens* 'Purple Sensation' OW	E–M Amaryllis. Purple spheres of small, shallow blossoms.	**Perennial** Hardy to zone 3. Deer resistant.	**2 ft (60 cm)** Mulch in cold zones. Suitable for pots. See blue globe allium.	**BB, HB, SB** See blue globe allium. Bumblebee and butterfly plant.
Aster, 'Mönch' *Aster × frikartii* 'Mönch' OW	M–L Aster. Lavender-blue, daisy-like flowers. **Group with:** Goldenrod, heliopsis, pincushion flower.	**Perennial** Hardy to zone 5. Deer and drought resistant.	**2–4 ft (60–120 cm)** Sun, well-drained soil. Pinch back May or June to encourage buds. Divide biennially. Suitable for fall border, drifts, bee pasture. Allow to self-seed.	**BB, HB, SB, BI** Cuckoo, leafcutter, long-horned, mining, green sweat, small carpenter bees. Key fall forage for long- and short-tongued bees
Aster, Stokes' *Stokesia laevis* 'Blue Danube' **NW:** Southeastern US Atlantic coast	M–L Aster. Purple to blue, dandelion-like flowers, some with white centres.	**Perennial** Hardy to zone 3. Deer and drought resistant.	**2 ft (60 cm)** Sun in average to fertile, moist, well-drained soil. Dislikes water-logged soils high in clay. Self-seeding. Mulch for winter.	**BB, HB, SB, BI** Accessible source of nectar for many bees and butterflies.
Bellflower *Campanula carpatica* 'Deep Blue Clips' OW	M Bellflower. Masses of purple-blue, bell-shaped flowers. **Group with:** Coreopsis, speedwell.	**Perennial** Hardy to zone 3. Deer resistant.	**1 ft (30 cm)** Partial shade in average, well-drained soil. Sow in spring with light to germinate, propagate by cutting or division.	**BB, HB, SB** Attracts leafcutter, mason, sweat bees. Cut back after bloom for second bloom for the bees.
Blanketflower, 'Burgundy' *Gaillardia × grandiflora* 'Burgundy' **NW:** Canada, western US	M–L Aster. Daisy-like, burgundy petals, red-brown centre. **Group with:** Purple coneflower, snapdragon.	**Perennial** Hardy to zone 3. Deer and drought resistant.	**2 ft (60 cm)** Sun, dry to medium sand to well-drained loam. Sow early spring or divide. Prune to 6 in (15 cm) after blooming in fall.	**BB, HB, SB, BI** Leafcutter, long-horned, small carpenter, sweat bees. Deadhead often for continual bloom.
Bugloss, Italian *Anchusa azurea* OW	E–M Borage. Clusters of sky-blue, shallow flowers, long nectar tube. Edible herb. *May irritate skin.* **Group with:** Borage, columbine, indigo, lupin.	**Biennial or Short-lived Perennial** Hardy to zone 5. Deer and drought resistant.	**40 in (1 m)** Sun, well-drained soil. Start seed 12 weeks before last frost. Long taproot. Deadhead if you do not want it to self-seed. *A. azurea* naturalized in BC. 'Dropmore Blue' developed in Manitoba in 1905.	**BB, HB, SB** Mostly nectar. Honey plant. Classic European bee plant. Significant bumblebee plant. Cut back for second bloom for the bees.
Catnip or Catmint, Faassen's *Nepeta × faassenii* OW	M–L Mint. Spike-like whorls of light-purple, bilabial blossoms with dark-purple nectar guides. **Group with:** Sage, sea holly.	**Perennia** Hardy to zone 3. Deer and drought resistant.	**1 ft (30 cm)** Hot, sunny sites with well-drained soil. Water during dry season to encourage nectar. Cut to ground after blossoms fade to encourage second bloom. Essential bee plant.	**BB, HB, SB** Honey potential up to 180–445 lbs/ac. A good plant to monitor bee species. Leafcutter, long-horned, mason, mining and small carpenter bees.

*Each pound per acre is equal to approximately 1.1 kilograms per hectare.

NAME AND NATIVE REGIONS	PLANT FAMILY, BLOOM, FOLIAGE AND COMPANIONS	HARDINESS	MAXIMUM HEIGHT AND GARDEN NOTES	BENEFITS TO BEES*
Cat Thyme *Teucrium marum* **OW**	M–L Mint. Purple with silver foliage. See germander species.	**Perennial** Hardy to zone 5. Deer resistant.	40 in (1 m) Some cats love this more than catnip. See germander species.	**BB, HB, SB, BI** Long-tongued bees reap nectar rewards. See germander species.
Crocus, Dutch *Crocus vernus* **OW**	E Iris. Petals in shades of yellow, white and purple, often with showy nectar guides. **Group with:** Nodding onion, Siberian squill.	**Perennial** Hardy to zone 3. Deer resistant.	4–6 in (10–15 cm) Sun to part shade. Well-drained soil, moderate moisture. Plant in fall, 3 times as deep as bulb height. Will naturalize in lawns. Divide every 3–4 years.	**BB, HB** If the weather is warm in early spring, bees seek out crocus blooms in droves.
Coreopsis, Threadleaf *Coreopsis verticillata* 'Moonbeam' **NW:** OT, QC, east-central US	M–L Aster. Daisy-like, pale-yellow flowers. **Group with:** Anise hyssop, *Campanula* spp., hardy geranium.	**Perennial** Hardy to zone 3. Deer resistant.	40 in (1 m) Sun to dappled shade in average to poor, well-drained soil. Sow spring or fall. Propagate by cuttings or division. Deadhead regularly.	**BB, HB, SB, BI** Bees of all stripes will visit coreopsis for nectar and pollen, as long as it's in a warm location, even if in shade.
Dahlia, 'Bishop of Llandaff' *Dahlia* 'Bishop of Llandaff' **NW:** Mexico, Guatemala	M–L Aster. Scarlet flowers fade to coral in sun. **Group with:** Pincushion flower, sunflower.	**Frost-tender Perennial** Hardy to zone 8+. Deer resistant.	40 in (1 m) Sun in fertile, humus-rich soil. Lift tubers after first hard frost, store in dry, cool, frost-free place until spring. Plant 4 in (10 cm) deep.	**BB, HB, SB, BI** Good late-season gap-filler for many bees. Avoid dahlias with double blossoms; choose singles for bees.
Elecampane *Inula helenium* **OW**	M–L Aster. Daisy-shaped yellow flowers with large flat centres. **Group with:** Scarlet bee balm, vervain.	**Perennial** Hardy to zone 3.	80–120 in (1–3 m) Full to partial sun in moist, well-drained soil. Start indoors or in cold frame spring or fall. Self-seeding. Border, meadow.	**BB, HB, SB, BI** Classic shallow-bowl open-access plant. Velvety centres perfect landing pads for bees, butterflies.
Fleabane, Philadelphia *Erigeron philadelphicus* **NW:** Canada; US except AK, UT, AZ	M–L Aster. Daisy-like, blue, pink or mauve flowers with yellow centres. Choose varieties with large centres.	**Annual or Biennial** Hardy to zone 3.	40 in (1 m) Sun to part shade in fertile, well-drained soil. Cut back after flowering. Shorter species perfect for rock garden.	**BB, HB, SB, BI** Cuckoo, mason, mining, small carpenter, sweat, masked bees.
Fleeceflower *Persicaria bistorta* **OW**	M–L Buckwheat. Showy spikes of small pink or red flowers.	**Perennial** Hardy to zone 3.	3 ft (90 cm) Sun to partial shade in moist, rich soil. Deadhead to encourage bloom.	**BB, HB, SB** Honey potential (*Persicaria* spp.): up to 90–180 lbs/ac. Nectar and pollen.
Fleeceflower, Mountain *Persicaria amplexicaulis* 'Firetail' **OW**	M–L Buckwheat. See fleeceflower. **Group with:** Blazing star, fall sedum.	**Perennial** Hardy to zone 3.	40 in (1 m) Suitable for mixed border. 'Alba' has white blossoms. See fleeceflower. Long-blooming, low-maintenance. Spreading, but not aggressive.	**BB, HB, SB** See fleeceflower. Important late-season plant for honeybees and bumblebee queens.
Fleeceflower, White *Persicaria polymorpha* **OW**	E–M Buckwheat. Giant, fluffy spikes with masses of white blooms. **Group with:** White coneflower.	**Perennial** Hardy to zone 3.	4 ft (1.2 m) Sun in moist, fertile soil. Protect from wind. Divide in spring. May need staking at first. Dies down in winter. Hedgerow, feature, back of border.	**BB, HB, SB** See fleeceflower. Significant butterfly plant. Birds eat the seeds in fall.

NAME AND NATIVE REGIONS	PLANT FAMILY, BLOOM, FOLIAGE AND COMPANIONS	HARDINESS	MAXIMUM HEIGHT AND GARDEN NOTES	BENEFITS TO BEES*
Geranium, 'Raven', a.k.a. Dusky Cranesbill *Geranium phaeum* 'Raven' OW	M–L Geranium. Dark-burgundy bloom, white centre; often marked with nectar guides. **Group with:** Masterwort.	**Perennial** Hardy to zone 3. Deer and drought resistant.	30 in (76 cm) Dappled shade, adaptable. Propagated by division. Suitable for shallow soil, shade; below trees, shrubs, eaves. Spreads quickly even in dry spots.	BB, HB, SB, BI Cuckoo, long-horned, mason, mining, sweat bees. Specialist pollinator: *Andrena distans.*
Geranium, 'Rozanne' *Geranium* 'Rozanne' OW	M–L Geranium. Mauve or blue. See 'Raven' geranium.	**Perennial** Hardy to zone 3. Deer and drought resistant.	20 in (50 cm) See 'Raven' geranium.	BB, HB, SB, BI See 'Raven' geranium.
Germander, Creeping *Teucrium aroanium* OW	M–L Mint. Pink flowers with large lower lips, no upper lip. **Group with:** Sedum.	**Perennial** Hardy to zone 5. Deer resistant.	3 in (7.5 cm) Sun in well-drained, moderate to dry, average to poor soil, including sandy, acidic sites. Pinch back after flowering.	BB, HB, SB, BI Long-tongued bees reap nectar rewards. Stamens give bees a pollen shower.
Germander, Shrubby *Teucrium fruticans* OW	M–L Mint. Pale-blue flowers. See creeping germander. **Group with:** Lavender, Russian sage.	**Perennial** Hardy to zone 5. Deer resistant.	40 in (1 m) Propagate by division or cuttings. Suitable for rock garden. Shelter plants in zone 5. See creeping germander.	BB, HB, SB, BI With all germanders, if nectar is running high, short-tongued bees can sip it up.
Germander, Wall *Teucrium chamaedrys* OW	M–L Mint. Pink. See germander species. **Group with:** Lavender, sage.	**Perennial** Hardy to zone 4 or 5. Deer resistant.	40 in (1 m) Good bones for garden in zones where lavender is tender. See germander species.	BB, HB, SB, BI See germander species.
Germander, Wood Sage *Teucrium scorodonia* OW	M Mint. Pale-yellow bloom, large lower lips, no upper lip. Spreads aggressively.	**Perennial** Hardy to zone 7. Deer and drought resistant.	1–2 ft (30–60 cm) Moderate shade to sun. Plant early spring or late fall. Barely cover seeds. See creeping germander.	BB, HB, SB, BI Adored by wool carder bees. See creeping and shrubby germander.
Globe Thistle *Echinops ritro* OW	M–L Aster. Globes of silvery blue flowers in second year. **Group with:** Russian sage and sea holly for a drought-tolerant drift of purple/blue.	**Perennial** Hardy to zone 3. Deer and drought resistant.	40 in (1 m) Sun in well-drained soil. Start seeds inside early spring: need light to germinate. Plant out once soil is warm. Long taproot. Propagate from sideshoots. Stake if needed.	BB, HB, SB, BI Honey plant. One of the most photogenic bee plants, attracting bees of all stripes. On sunny days, visited by clouds of insects.
Goldenrod, Sweet *Solidago odora* NW: Southern and eastern US	M–L Aster. Clusters of bright yellow flowers. Foliage has an anise scent.	**Perennial** Hardy to zone 4. Deer and drought resistant.	2–5 ft (60–150 cm) Sun to part shade in well-drained soil. Cold stratify seeds or plant in the fall, barely covering.	BB, HB, SB, BI Honey potential (*Solidago* spp): 25–50 lbs/ac. Large carpenter, mining, sweat bees.
Heather, Ling or Scottish *Calluna vulgaris* OW	E–M–L Heather. White or pink bell-shaped blooms in winter to early spring or summer to fall. **Group with:** Wild onions and sea thrift.	**Perennial** Hardy to zone 5. Deer resistant.	1–3 ft (30–90 cm) Sun to partial shade in moist, acidic, well-drained soil. Suitable for edging, hedgerow understory. Some heather is more attractive to bees than others—consult a nursery.	BB, HB, SB Honey potential: 100–200 lbs/ac; slow to granulate; gel-like consistency makes it unsuitable for winter use by honeybees. Attracts small bees.

NAME AND NATIVE REGIONS	PLANT FAMILY, BLOOM, FOLIAGE AND COMPANIONS	HARDINESS	MAXIMUM HEIGHT AND GARDEN NOTES	BENEFITS TO BEES*
Heliopsis, a.k.a. Sweet Oxeye, False Sunflower *Heliopsis helianthoides* 'Summer Sun' **NW:** Southwestern and central Canada; central, eastern US	**M–L** Aster. Daisy-like blossoms with yellow petals. This cultivar better behaved than its wild form, which can be thuggish. Blooms 10–13 weeks. **Group with:** Pincushion flower, zinnia.	**Perennial** Hardy to zone 3. Drought resistant.	**40–80 in (1–2 m)** Full to partial sun in clay, sand or loam, in average to dry conditions. Easy to start from seed. Divide spring or fall. Deadhead flowers. Suitable for bee pasture, hedgerow, woodland-edge garden.	**BB, HB, SB, BI** Classic shallow-bowl-shaped yellow flowers attract bumblebees, digger, cuckoo, mining, small carpenter, sweat bees. Deadhead so blooms for bees continue.
Hollyhock *Alcea rosea* **OW**	**M–L** Mallow. Shallow bowl-shaped flowers in many colours. Flowers and leaves are edible. Avoid double blossoms.	**Biennial or Perennial** Hardy to zone 2.	**40–80 in (1–2 m)** Sunny, well-drained soil. Start seeds indoors. Hollyhock likes cool roots and warm heads. Prone to rust. Suitable for the back of border.	**BB, HB** Large pollen grains. Sugar concentration 34 percent. An important late-season pollen plant, with nectar rewards too.
Honeywort *Cerinthe major* **OW**	**M** Borage. Blue bracts and purple flowers. **Group with:** Borage, bugloss.	**Frost-tender Perennial** Hardy to zone 8.	**17 in (43 cm)** Sun in a warm, sheltered location in well-drained soil. Easily grown from pre-soaked seed.	**BB** Bumblebees love this nectar-rich flower. Look for signs of nectar robbing.
Leadwort, a.k.a. Blue Plumbago *Ceratostigma plumbaginoides* **OW**	**M–L** Leadwort. Deep-blue flowers. Bronze foliage in winter. **Group with:** Goldenrod, sedum.	**Perennial** Hardy to zone 5. Deer and drought resistant.	**1 ft (30 cm)** Sunny, well-drained soil. Does not like wet feet in winter. Propagate by division and cuttings. Suitable for ground cover.	**BB** Important plant for new fall bumblebee queens. Late-season forage for long-tongued bees.
Lungwort *Pulmonaria officinalis* **OW**	**E** Borage. Pink bell-shaped blossoms turn blue as they age. Foliage has distinct silver spots.	**Perennial** Hardy to zone 3. Deer resistant.	**1 ft (30 cm)** Shade. Needs winter protection. Keep moist but avoid wetting leaves to prevent powdery mildew. Cut back after blooming.	**BB, SB** Early-spring bee plant. Varieties that bloom later tend to attract more bees.
Mallow, Zebra *Malva sylvestris* 'Zebrina' **OW**	**M–L** Mallow. Pink to purple flowers with showy nectar guides. **Group with:** Hollyhock, pink bellflower, snapdragon.	**Biennial or Perennial** Hardy to zone 4.	**40 in (1 m)** Sun to dappled shade in rich, moist soil. Easy to grow from seed. Start indoors in early spring or direct-sow fall. Edible flowers and leaves.	**BB, HB, SB, BI** Cuckoo, leafcutter, mining, sweat. Bees become covered in large, pink pollen grains. Butterfly plant.
Masterwort *Astrantia major* **OW**	**E–M–L** Carrot. Flowers composed of tiny white, pink or burgundy florets. **Group with:** Hardy geranium, snowberry.	**Perennial** Hardy to zone 4. Deer resistant.	**40 in (1 m)** Sun to partial shade in well-drained humus-rich soil. Will tolerate drier conditions in dappled shade. Good border and hedgerow understory.	**BB, HB, SB, BI** Florets attract a wide variety of bees. Important long-blooming, shade-tolerant plant for bees.
Obedient Plant, 'Miss Manners' *Physostegia virginiana* 'Miss Manners' **NW:** Eastern Canada and US	**M–L** Mint. Tall stalks of pink snapdragon-like flowers. **Group with:** Snapdragon.	**Perennial** Hardy to zone 2. Deer resistant.	**25 in (64 cm)** Sun or partial shade in well-drained soil. Fantastic for cottage garden and perennial border.	**BB, SB** Bumblebees climb right inside to gather nectar. Nectar robbing by large carpenter bees.
Penstemon, Shrubby *Penstemon fruticosus* 'Purple Haze' **NW:** BC, northwestern US	**E–M** Plantain. Purple-blue bilabial flowers, long corollas with violet nectar guides. 'Purple Haze' bred for more flowers than native variety.	**Perennial** Hardy to zone 4. Deer and drought resistant.	**6 in (15 cm)** Sun in sandy, well-drained soil. Long-tongued bees seek nectar/pollen; small bees gather pollen from stamens.	**BB, HB, SB, BI** Digger, mason, mining, large leafcutter, small carpenter, sweat, wool carder bees.

NAME AND NATIVE REGIONS	PLANT FAMILY, BLOOM, FOLIAGE AND COMPANIONS	HARDINESS	MAXIMUM HEIGHT AND GARDEN NOTES	BENEFITS TO BEES*
Pincushion Flower *Scabiosa columbaria* 'Butterfly Blue' **OW**	M–L Honeysuckle. Large lavender-blue flowers. **Group with:** Dahlia and tall vervain.	Perennial Hardy to zone 4.	1 ft (30 cm) Sun, average soil. Deadhead to encourage blooming. Some varieties self-seed. Grow cultivars by division.	BB, HB, SB, BI Leafcutter, mining bees and more. Bumblebees sleep on flower heads. Photogenic bee-garden essential.
Poppy, Icelandic *Papaver nudicaule* **NW:** AK, YT	E–M Poppy. Cup-shaped, bright-coloured flowers, copious stamens. Avoid double blossoms.	Perennial Hardy to zone 1. Deer and drought resistant.	2 ft (60 cm) Light is required for germination. Start in early spring. See Welsh poppy.	BB, HB, SB Flowers loved by small bees that swim in the stamens collecting pollen.
Poppy, Welsh *Meconopsis cambrica* **OW**	E–M Poppy. See Icelandic poppy. **Group with:** California poppy and tidy tips.	Perennial Hardy to zone 6. Deer and drought resistant.	17 in (43 cm) Full to dappled sun in well-drained soil. Will tolerate many soils, but best in rich, moist soil. Self-seeding.	BB, HB, SB See Icelandic poppy. Collect and scatter seeds to grow more pollen-rich blooms for bees.
Sage, Lilac *Salvia verticillata* 'Purple Rain' **OW**	M–L Mint. Spikes of purple bilabial flowers. **Group with:** Lavender, thyme.	Perennial Hardy to zone 4. Drought resistant.	2 ft (60 cm) Sun, well-drained soil. Deadhead for second bloom. Suitable for pots.	BB, HB, SB Important source of nectar in late summer–fall for bees, butterflies.
Sage, Russian *Perovskia atriplicifolia* **OW**	M–L Mint. Elongated spears of purple bilabial flowers. Silver-grey foliage. **Group with:** Globe thistle, sea holly.	Perennial Hardy to zone 3. Deer and drought resistant.	40 in (1 m) Sun, well-drained soil. Deadhead to extend bloom; cut right down after blooming. Dies back in winter. Suitable for hedgerow, boulevard.	BB, HB, SB Mostly nectar. Leafcutter, sweat bees. Essential late-season bee plant. Bumblebees sleep overnight on flowers.
Sage, Woodland *Salvia nemorosa* 'Mainacht' **OW**	M–L Mint. Spikes of purple bilabial flowers. **Group with:** Lavender, wall germander.	Perennial Hardy to zone 3. Deer and drought resistant.	2 ft (60 cm) Sun, well-drained soil. Propagate by root division or cuttings. Suitable for border, boulevard.	BB, HB, SB Honey potential: 180–445 lbs/ac. Important late-summer to fall nectar. Essential bee plant.
Sea Holly, Blue *Eryngium alpinum* **OW**	M–L Carrot. Prickly foliage and thumb-shaped flowers are silver to deep blue.	Perennial Hardy to zone 2. Deer and drought resistant.	2 ft (60 cm) Sun in well-drained, moderate to poor soil with grit. Lime tolerant. Long taproot: don't transplant.	BB, HB, SB, BI Essential drought-tolerant bee plant. Leafcutter, mining bees and more.
Sea Thrift *Armeria maritima* **OW/NW:** Canada except AB, NB, NS; US Pacific coast, CO, PA	E–M–L Leadwort. Globes of bright-pink blooms. **Group with:** Pastel shades of yarrow, wild onions.	Perennial Hardy to zone 2. Deer resistant. Salt tolerant.	1 ft (30 cm) Sun in poor, well-drained soil, under dry conditions. Don't prune. Slow-spreading. Suitable for border, rock, rooftop gardens, no-mow zones.	BB, HB, SB Mason, mining bees and more. Nectar. Deadhead to encourage blooms for bees.
Sedum, a.k.a. Stonecrop *Hylotelephium spectabile* **OW**	M–L Stonecrop. Masses of shallow pink blossoms. Foliage green to burgundy. **Group with:** Brown-eyed Susan, purple coneflower.	Perennial Hardy to zone 5. Drought resistant.	17 in (43 cm) Sunny location in fertile, well-drained soil. Propagate by division in spring or softwood cuttings in early summer. Border or boulevard garden.	BB, HB, SB Late-season source of pollen and nectar. Bumblebees and honeybees frantically scramble over blossoms to feed on nectar.
Sedum, 'Purple Emperor' *Sedum telephium* 'Purple Emperor' **OW**	M–L Stonecrop. Masses of shallow pink blossoms. Dark-purple foliage. **Group with:** Aster, Russian sage.	Perennial Hardy to zone 2.	20 in (50 cm) Sunny location in fertile, well-drained soil. Dark foliage looks dramatic against silvery blue plants.	BB, HB, SB Top dependable late-season plant for bees. Rafts of shallow blossoms make foraging efficient.

NAME AND NATIVE REGIONS	PLANT FAMILY, BLOOM, FOLIAGE AND COMPANIONS	HARDINESS	MAXIMUM HEIGHT AND GARDEN NOTES	BENEFITS TO BEES*
Sedum, White, a.k.a. Houseleek *Sedum album* **OW**	M Stonecrop. White or pink to red star-shaped blossoms.	**Perennial** Hardy to zone 3. Drought resistant.	6 in (15 cm) Sun or partial shade in rocky, well-drained soil. Propagate by division or cuttings. Suitable for pots.	BB, HB, SB, BI Easy-access nectar. Significant bumblebee and butterfly plant.
Snapdragon *Antirrhinum majus* **OW**	M–L Plantain. Velvety bilabial flowers in variety of colours. Toxic. **Group with:** Toadflax, turtlehead.	**Frost-tender Perennial** Hardy to zone 5.	40 in (1 m) Sun to partial shade in rich soil with neutral pH. Start indoors 8 weeks before setting out after soil warms. Deadhead. Cut back if plant is fading.	BB, SB, BI Bumblebees love this plant and have the heft to trip blossoms for nectar. Look for signs of nectar robbing.
Sneezeweed *Helenium autumnale* **NW:** Most of Canada and US	M–L Aster. Native species has yellow petals but cultivars range from yellow to red with centres ringed with yellow pollen. Leaves, flowers, seeds mildly toxic.	**Perennial** Hardy to zone 4.	5 ft (1.5 m) Prefers sun in moist, well-drained soil. Cut plants down after blooming. Divide every few years, depending on cultivar. May need staking.	BB, HB, SB, BI Helps honeybees build up stores for winter. A late-season source of pollen and nectar. Bee-garden essential.
Speedwell, Spike *Veronica spicata* **OW**	E–M Plantain. Spikes of blue to purple flowers, protruding stamens, short nectar tubes. 'Sunny Border Blue' is a cultivar popular with bees.	**Perennial** Hardy to zone 4.	8 in (20 cm) Sun to dappled shade in well-drained soil. Long-blooming, low maintenance. Deadhead.	BB, HB, SB, BI No bee garden should be bereft of this nectar and pollen dynamo. Spiked varieties especially attractive to bees.
Speedwell, Tall *Veronica longifolia* **OW**	E–M Plantain. See spiked speedwell. Cultivars available in different colours, including white and pink.	**Perennial** Hardy to zone 4.	2 ft (60 cm) See spike speedwell. Avoid planting invasive *V. arvensis, V. beccabunga* and *V. repens.*	BB, HB, SB, BI See spiked speedwell. *Veronica* spp. good nectar and pollen producers.
Squill, Siberian *Scilla siberica* **OW**	E Asparagus. Nodding blue flowers. **Group with:** Will grow below pine and black walnut.	**Perennial** Hardy to zone 4.	8 in (20 cm) Sun to part shade in well-drained soil. Plant bulbs 3 in (7.5 cm) deep and apart. Suitable for rock garden, naturalizing.	BB, HB, SB Nectar and blue pollen. Important early-season bee plant for solitary bees and bumblebees.
Stonecrop, Caucasian *Sedum spurium* **OW**	M Stonecrop. White or pink to red star-shaped blossoms.	**Perennial** Hardy to zone 3. Drought resistant.	6 in (15 cm) Sun or partial shade in rocky, well-drained soil. Division or cuttings.	BB, HB, SB, BI Easy-access bee plant. Bumblebees sleep on the flowers.
Sweet Pea, Perennial *Lathyrus latifolius* **OW**	M–L Legume. Pink flowers trip-pollinated by long-tongued bees. Flowers and young pods edible in small amounts. However, annual sweet peas are highly toxic. **Group with:** Sweet peas like warm heads/cool roots, so plant low-growers around them.	**Perennial** Hardy to zone 5. Drought resistant.	80 in (2 m) Sun in normal to sandy, well-drained soil. Scarify seeds with sandpaper, pre-soak. Start indoors before last frost; transplant once soil is workable. Keep moist, use netting to protect from birds. Needs trellis, air circulation.	BB, SB Bumblebees and leafcutter bees drink the copious nectar and harvest bright-yellow pollen. Water in morning to encourage nectar production.
Toadflax, Purple *Linaria purpurea* **OW**	E–M–L Plantain. Spikes of bilabial purple, pink or white flowers. Toxic. **Group with:** Cranesbill geranium, penstemon.	**Perennial** Hardy to zone 5. Deer resistant.	40 in (1 m) Sun, well-drained soil. Self-seeding. If looking for something less weedy, try *L. maroccana.* Avoid invasive *L. vulgaris.*	BB, HB, SB, Small carpenter, sweat, wool carder bees. Bees trip the lips as they seek nectar, dusting their bellies with pollen.

NAME AND NATIVE REGIONS	PLANT FAMILY, BLOOM, FOLIAGE AND COMPANIONS	HARDINESS	MAXIMUM HEIGHT AND GARDEN NOTES	BENEFITS TO BEES*
Vervain, a.k.a. Tall Verbena *Verbena bonariensis* **NW:** South America	**M–L** Verbena. Clusters of mauve blossoms at the end of long, airy stems. **Group with:** Pincushion flower, zinnia.	**Perennial** Hardy to zone 7. Drought resistant.	**80 in (2 m)** Sun, moist, well-drained soil. Divide, take cuttings. Sow spring or fall. Start indoors 8 weeks before last frost. Self-seeding.	**BB, HB, SB** Loved by long-tongued bees, hummingbirds, butterflies.
Yarrow *Achillea millefolium* **OW/NW:** Canada, continental US		**Perennial** Hard to zone 3–4, depending on cultivar. Deer and drought resistant.	**2 ft (60 cm)** Sun, part shade in sand, loam or clay. Dry to medium conditions. Seed needs light to germinate. Also spreads by rhizomes. Great in pots.	**BB, HB, SB, BI** Essential gap-filler for short-tongued bees, including mining and sweat bees. Wool carder bees gather hairs on stems to line nests.

*Each pound per acre is equal to approximately 1.1 kilograms per hectare.

This honeybee has been collecting yellow pollen from another plant before feasting on drumstick allium, which has bright purple pollen.

Growing to Love Bees

CHAPTER TEN

*E*very Victory Garden has a story, a narrative uniting families, neighbours and colleagues. In World War II, children, teens, women and seniors worked side by side in schools, yards and fields to grow food for themselves and to share with others. Victory Gardening was a community-building, morale-boosting exercise. And today, at a time when technology is luring all of us into an insular indoor world, it's more important than ever to get out and get planting flowers for bees with family and friends, old and young, making it fun.

WHY IT'S GOOD TO GET YOUR HANDS IN THE SOIL

During one of my bee-gardening workshops, I handed a six-year-old boy a seedling to plant in the school plot. He admired the plant, but wasn't so sure about the soil around its roots. "Can you put this in the garden for me, because I don't like to get my hands dirty?" "Okay," I said, "put the plant down and take your pinky finger and put it in the soil. Just take it one finger at a time." The next thing you know, he's happily planting carrot seeds with his bare hands. A few months later, he was pulling orange, purple and yellow carrots out of the soil, eager to dig for this buried treasure.

When a child asks me to put a plant in the soil because he doesn't want his hands to get dirty, I feel a deep, poignant sadness. When I put my hands in the soil, something magical happens: my body connects with the soil and remembers its purpose . . . *to garden until I can garden no more.*

Opposite: Students can create recycled bicycle-wheel trellises for the school garden. Scarlet runner bean blossoms have magnetic appeal for bumblebees and hummingbirds.

> ### *Watch Them Grow*
>
> To share the miracle of seeds, provide a child with a brown paper coffee filter (or paper towel) and help them dampen it. Give them an assortment of seeds to arrange on the paper filter in whatever pattern they like. Gently slide the seedy result into a plastic bag, seal it, and over the next few days watch the seeds sprout.

TEACH CHILDREN TO BE SEED GLEANERS

Saving seeds and growing plants from seeds are essential skills for supporting bees. A little girl in one of my classes asked me, "Can plants make more than one seed?" "Yes, one plant can produce up to a million seeds, when they've been pollinated by bees." Her eyes widened and I saw the seed of understanding fall on the fertile ground of her consciousness. It never fails that when I take seeds into a classroom, they work their magic on children and adults. The potent metaphors of a life in waiting, the life cycle, a promise of new life, are all contained in a seed. On warm summer and fall afternoons, help kids gather mature seed pods and seed heads to dry in paper bags for springtime shaking and scattering.

LET THE KIDS GO WILD FOR BEES

California has an incredibly diverse population of bees, especially in some of the arid regions. The flowers have such irresistible names: baby blue eyes, poached egg flower, tidy tips and Chinese houses are just a few. Many of these wildflowers thrive on neglect, making them perfect self-seeders to give to kids to sprinkle in gardens or pots as gifts to bees.

PLANTING PLAN

Rocky Mountain Bee-safari Park: A Children's Bee Garden

Inspired by summer camps in Alberta's Rocky Mountains, this garden pulses with the possibility of adventure. Children of all ages can snap bee-safari photos while nibbling on foraged kale flowers and blueberries. Canoes are repurposed for garden plots, carrying loads of carrots and pumpkins amidst a swirl of blue bee-meadow flowers. Include a sturdy boat for impromptu picnics and hours of imaginary trips paddling down the Bow River. Use your imagination to repurpose other items as containers for planting—from bathtubs and bed frames to old wagon wheels and cars. (Just be sure there is good drainage so that they don't become more suitable for fish than bees; drill as needed.) The alpine bee hotel provides deluxe accommodation for local pollinators and allows hours of fun watching them provision their nests. This is the perfect project for a scout club, community centre or school. See opposite page for a sample Rocky Mountain bee-safari garden plan.

TALES OF A BEE SAFARI

On a muggy August afternoon, I arrive at the Cedar Cottage Neighbourhood House to do a bee safari with a group of six- to eight-year-olds. "Do you know how to act around bees?" I ask.

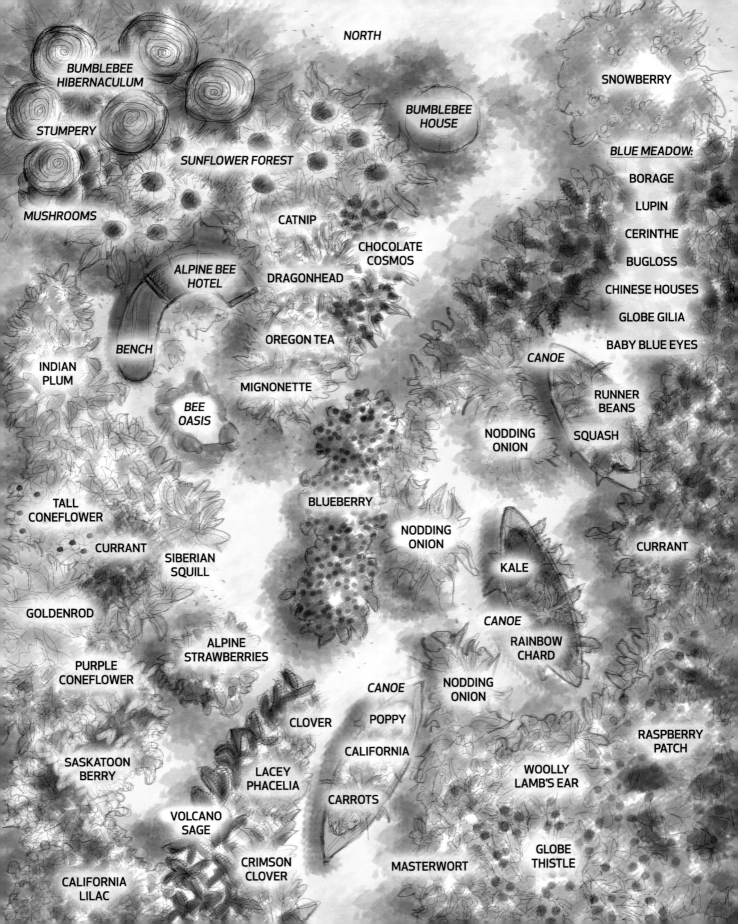

Nose Twisters and Bee Popsicles

Nasturtium and calendula are colourful edible flowers for bees, with nice big seeds that are easy for kids to plant.

Dried nasturtium seeds were used as a pepper substitute in World War II; "nasturtium" means "nose twister," likely inspired by their spicy flavour. The edible flowers are packed with vitamin C. Hummingbirds and long-tongued bumblebees love them, so plant the bright orange and red flowers with long nectar spurs and avoid nasturtiums without spurs because they lack rewards for bees and birds.

A generous, long-blooming flower with popsicle-orange blossoms, calendula is one of the easiest herbs to grow from seed, making it perfect for school gardens. Calendula's petals are edible and can be sprinkled on crackers with creamy cheese for a healthy snack; they are also are safe and gentle when used in homemade skin creams for babies and adults. Grow this pretty plant between rows of vegetables to preserve moisture in the soil and suppress weeds.

Sidebar: Calendula blooms profusely at the City Farmer children's garden created by resident "bug lady" Maria Keating.
Top: Tucked in a back alley, this canoe garden is surprisingly productive, with peas and beans growing up the springs from an old mattress.

"Don't do this," one girl says, as she runs around flapping her hands and screeching.

"Yes, that's exactly what you *don't* do," I say. "You need to take a deep breath and move very slowly. Do you know what's in here?" I ask, showing them a squash blossom I have pinched closed at the top.

"Is it a bee?" another girl asks.

"Yes, and I'm going to see if she'll crawl on my hand." I watch a frisson of revulsion shiver through her shoulders.

"I'm scared," she says.

I open the top of the blossom and the bee flies up and away. "You see, she doesn't want to bother me, she just wants to pollinate this zucchini." I explain how bees take pollen from the male blossom and put in on a female blossom so it can produce a squash like the ones we see developing on the plant.

Perched above the cars on a boulevard tree, a paper wasp nest hangs from a tree. "It's a bee nest!" the children shout and point upwards. I explain that this is *not* a home for bees. Wasps chew up wood to create a paper pulp that they form into their nest. While we watch bees foraging in the lavender, I explain that the blond bumblebees are males and they don't sting. "You can even catch them in your hands. Do you think I should do it?" They dare me and I catch a little blondie that buzzes in my cupped hands and then flies away.

"Do it again! Do it again!" they cry. Next, we see a wasp chomping down on a bee caught in a spider's web.

I take out the remainder of the dead honeybee to show them and they take turns touching her.

"Can you break that spider web so it doesn't catch any more bees?" asks the girl who was formerly afraid of bees. We form a circle and do a honeybee dance before the kids settle down for their snack.

HOW TO WIN FRIENDS AND INFLUENCE BEES

Whether it's a bat, a bear or a bee, we all live with the risks of having wild creatures among us and need to learn appropriate safety. Don't allow the fear of being stung to get in the way of enjoying outdoor fun. Remember that the majority of bee species are solitary and have a weak sting or no sting at all. The majority of what people think of as bee stings are actually from wasps.

The city of Brussels has declared May 1 as International Sunflower Guerrilla Gardening Day. Imagine pollinator paths created by back alleys and ditches brimming with sunflowers.

A Clean Bee: What Do You Do with a Stranded Queen?

Imagine my surprise when I opened up my washing machine in the basement to discover a rather large queen bumblebee resting in the lint trap. She had made a nest in an abandoned mouse burrow that led into the basement of our old house. Instead of heading back outside, Her Majesty sought the warmth inside and ended up in the washing machine. If you find a stranded bumblebee or honeybee, use a piece of paper to slip under her body and move her to a safe, sunny spot. Give her a few drops of honey or a 1:1 solution of sugar dissolved in water in an eye-dropper or spoon and send her on her way.

Opposite, clockwise from top left: Conservation biologist Erin Udal shows the pollinator hotel to campers at VanDusen Botanical Garden (GILLIAN DRAKE PHOTO). Create biodegradable sculptures to introduce children to the tactile pleasure of working with seeds. Performing as the Queen Bee, I teach children how to boogie like honeybees and perform the various jobs in the hive. **Above, right:** Globe gilia produces striking blue flowers with blue pollen that bees collect with a frenzied buzz.

Before Leaving the House

- ⚘ **IF YOU ARE ALLERGIC TO BEE STINGS**, wear long pants, long sleeves and a hat, and carry your EpiPen and cell phone. Rehearse using the pen with your family and friends *before* a bee sting happens: *Blue to the sky, orange to the thigh.*
- ⚘ **AVOID GOING BAREFOOT** if there are flowers full of bees in the lawn.
- ⚘ **BE FRAGRANCE-FREE.** Avoid using toiletries with heavy fragrances. Shower and wear fresh clothing, as sweat alerts bees to danger.

Honeybee Hive Safety

- ⚘ **WEAR LIGHT-COLOURED CLOTHING** around hives.
- ⚘ **TAKE A DEEP BREATH AND STAY CALM** and make slow, gentle movements.
- ⚘ **IF A BEE LANDS ON YOUR SKIN** to inspect a scent, remain still and it will peacefully leave once its curiosity has been satisfied. If you don't want to wait for it to leave, gently and slowly brush it away with a piece of paper, or ask someone else to brush it off you.

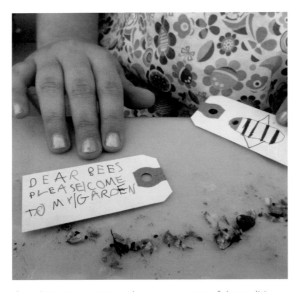

A student writes a personal message as part of the tradition of "telling the bees."

Artist Jasna Guy compares the pollen on fireweed stamens with the color samples created by Dorothy Hodges in her book *The Pollen Loads of the Honey Bee.* JASNA GUY PHOTO

- **NEVER SWAT AT A BEE.** If you kill a bee or wasp, it will send off alarm pheromones to attract other hive members.
- **DO NOT STAND DIRECTLY IN FRONT OF THE HIVE** where bees laden with nectar and honey are making a bee-line to the entrance.
- **HONEYBEES GUARDING THEIR HIVE WILL GIVE YOU A WARNING,** bonking your forehead or buzzing about your ears to let you know you are too close.
- **IF A BEE IS BUZZING AROUND YOU** and won't let you alone, move into the shade or indoors.
- **IF THE HIVE IS OPEN,** wear a full beekeeping suit if possible.

Travel Wise
- **IF A BEE OR WASP FLIES INSIDE YOUR VEHICLE,** gently stop the car and let it out.
- **IF YOU ARE VACATIONING IN PLACES WITH AFRICANIZED HONEYBEES,** review the safety procedures for dealing with those bees *before* you leave.

Picnic Protection

It's not bees, but rather wasps that are the annoying culprits at picnics. You can let them share a bite of your salmon placed away from the picnic blanket, but follow these tips to avoid stings:
- **DRINK FROM A GLASS.** Wasps climb into soda cans without being noticed, and sting your lip or mouth when you take a drink.
- **KEEP YOUR GARBAGE SEALED,** especially in late summer when wasp populations are high.
- **IF YOU ACCIDENTALLY STEP ON A WASP NEST OR BUMP INTO ONE,** run and seek the nearest shelter. Protect your throat and face.

UNFRIENDING BEES

Occasionally you will be stung by accident: by sitting on a bee, or stepping on one in your bare feet. It hurts a lot and you will feel like unfriending bees. Don't take it personally. Bad stings happen to good people. Once the pain goes away, you'll be ready to forgive and refriend the bees. Female honeybees (apart from the queen) die after stinging mammals, so they really don't want to hurt you or themselves; it's just part of their "fight or flight" response. Other bees, like bumblebees, have a smooth stinger, so they can sting more than once, but they rarely do. Most bees can't sting at all and no male bees from any species can sting.

TELLING THE BEES

In many cultures, bees have been revered as spiritual messengers, carrying news back and forth between our world and the invisible world of our ancestors. When there was an important event in the village, it was integral to tell the bees. There is a folk saying:

> *Marriage, birth or burying,*
> *News across the seas,*
> *All your sad or merrying,*
> *You must tell the bees.*

THE ART OF LOVING BEES

Artist Jasna Guy and I discovered we both had an interest in a book by Dorothy Hodges called *The Pollen Loads of the Honeybee.* We made a date and met at the University of British Columbia library, where we settled in to explore the rare edition reverently in a quiet room. We opened it to a page of delicate squares of amber paper carefully mounted to the pages. These depict samples of hazel and other pollen the author scraped from the baskets on bees' legs, then painstakingly matched with watercolour. They were screenprinted to her specifications and glued into place by hand. Afterward, I couldn't help but lie awake at night wondering what Dorothy Hodges's life was like during World War II, tending her bees and methodically painting her pollen charts.

Giant Melons for Wedding Cakes

During World War II, candied citrus fruits could not be imported from Italy, so farmers in the Okanagan grew fields of giant gourds called "zucca melons." A single gourd could weigh as much as 127 pounds (58 kg), and the flesh was candied for fruitcakes, which were commonly served at weddings and Christmas celebrations. Originating from Africa, these gourds are pollinated by moths at night, so had to be hand-pollinated in Canadian fields after sunset.

VICTORY GARDEN CARE KITS FOR BEES

During World War II, women made care kits for soldiers, knitting thousands of pairs of socks and canning gallon after gallon of jam from their Victory Garden harvests. You, too, can make Victory Garden care kits, this time for bees, by packaging seeds for friends and neighbours to plant for the bees. Choose easy-to-grow annuals like buckwheat, calendula, crimson clover, flax, lacey phacelia and hairy vetch.

LET'S GET TOGETHER TO PLANT VICTORY GARDENS FOR BEES

In order to plant the sheer volume of flowers needed to save the bees, we have to channel the Victory Garden spirit and combine our efforts. One of the meanings of the word "bee" is a gathering where people come together to make or build something meaningful to all. While traditionally this has meant bees for knitting, canning, quilting and building barns, now is the time for "Bee-gardening Bees"! It's vitally important to work in community to plant interconnected pesticide-free pollinator habitats of all shapes and sizes to support bees of all stripes. It is also essential to protect and restore the wild habitat that is integral to native bees.

Every time you sit down to nourish yourself, remember the bees and the irreplaceable role they play in feeding us, keeping our farms and gardens filled with flowers and fruit. Become a bee guardian and play a heroic role in helping your community grow clean, safe and nutritious Victory Gardens for Bees.

WE CAN DO IT! WE CAN SAVE THE BEES

At the onset of World War II, there was doubt among the higher-ups in the Canadian War Office that ordinary citizens would know what to do with tools and seeds for backyard gardening.[58] But even though many Canadians had little or no experience, they used their "can-do" attitude to create abundant Victory Gardens.

In fact, the Victory Garden movement in Canada was more citizen-powered than that of any other allied country. Not only did Canadians send *tons* of food overseas to soldiers and civilians, but their own families ate better than ever before. History tells us it *is* possible to create big change with a few seeds and a ready spirit.

Gardeners everywhere, experienced or not, can take inspiration from the chutzpah of our ancestors and rally to make a difference. Dig for Victory! Grow and share your Victory Garden for Bees with friends, family and our beloved bees.

Above: In Washington State, the Noxious Weed Control Board is distributing free packages of bee-friendly wildflower seeds. Design by Drake Cooper, seeds by Ed Hume Seed Co. **Opposite, clockwise from top left:** A leafcutter bee forages among nasturtiums spilling from the edges of a raised vegetable bed.

UBC Farm has become a popular place for outdoor summer weddings, offering up bee-friendly bouquets from on-site flower gardens. Give children the opportunity to dress in bee regalia and "pollinate" the community garden.

ENDNOTES

1. Walsh, Bryan. "Going Green: Beepocalypse Now?" *TIME*. September 13, 2007. http://content.time.com/time/magazine/article/0,9171,1661683,00.html

2. Winston, Mark L. "Toxic Soup." The Hive. February 8, 2014. http://winstonhive.com/?p=191

3. Pearson, Gwen. "You're Worrying About the Wrong Bees." *WIRED*. April 29, 2015. http://www.wired.com/2015/04/youre-worrying-wrong-bees/

4. Moisset, Beatriz, and Stephen Buchmann. *Bee Basics: An Introduction to Our Native Bees.* USDA Forest Service and Pollinator Partnership. 6.

5. Goulson, Dave. "There Is No Plan Bee for When We Run out of Pollinators." *Financial Times*. November 8, 2013. http://www.ft.com/cms/s/0/a7ffe730-47a0-11e3-9398-00144feabdc0.html#axzz355QkJmkj

6. "Bee Wise" is a term first suggested to me by another artist who runs with the bees, Brenna Maag.

7. Before digging into any new areas on your property or a public boulevard, check that it is safe to do so by applying for any civic or county permissions required. Refer to your property records for buried wires, and irrigation or drainage infrastructure.

8. "Bio-plan" is a term coined by botanist and biochemist Diana Beresford-Kroeger to describe a balanced garden that a person uses to improve their quality of life and mend the health of the planet. http://dianasjourney.com/

9. Case, Elizabeth. "Insecticide Temporarily Banned by Oregon Department of Agriculture after 50,000 Bumblebees Die in Wilsonville." *Oregon Live*. June 27, 2013. http://www.oregonlive.com/environment/index.ssf/2013/06/state_agency_temporarily_bans.html

10. Elle, Elizabeth. Lecture, "Bees in the City." Vancouver Park Board. July 4, 2014.

11. Thomson, Janet, and Manmeet Ahluwalia. "What's Killing Canadian Honeybees?" *CBC News*. July 9, 2013. http://www.cbc.ca/news/canada/what-s-killing-canadian-honeybees-1.1312511

12. Ibid.

13. Leahy, Stephen. "Neonicotinoids: The New DDT." *Watershed Sentinel*. August 28, 2014. http://www.watershedsentinel.ca/content/neonicotinoids-new-ddt

14. Elle, Elizabeth. Lecture, "Bees in the City." Vancouver Park Board. July 4, 2014.

15. "Neonicotinoid Pesticides Are a Huge Risk—So Ban Is Welcome, Says EEA." *European Environment Agency*. May 2, 2013. http://www.eea.europa.eu/highlights/neonicotinoid-pesticides-are-a-huge

16. Atkins, Eric. "Ontario First in North America to Restrict Pesticides Blamed for Bee Decline." *Business News Network*. http://www.bnn.ca/News/2015/6/9/Ontario-first-in-North-America-to-restrict-pesticides-blamed-for-bee-decline.aspx

17. Theen, Andrew. "Portland Bans Use of Insecticides Believed to Be Harmful to Bees on City Property." *The Oregonian*. June 18, 2013. http://www.oregonlive.com/portland/index.ssf/2015/04/portland_bans_use_of_specific.html

18. Hightower, Steven. "Insectaries and IPM at Benziger Winery." *UC Master Gardener Program of Sonoma County*. http://ucanr.edu/sites/scmg/Feature_Articles/Insectaries_and_IPM_at_Benziger_Winery/

19. Keim, Brandon. "Beyond Black and Yellow: The Stunning Colors of America's Native Bees." *WIRED*. August 12, 2013. http://www.wired.com/2013/08/beautiful-bees/?pid=7213&viewall=true

20. Grissell, Eric. *Bees, Wasps, and Ants: The Indispensable Role of Hymenoptera in Gardens.* Portland, Oregon: Timber Press, 2010. 88–89.

21. Keim, Brandon. "Beyond Black and Yellow: The Stunning Colors of America's Native Bees." *WIRED*. August 12, 2013. http://www.wired.com/2013/08/beautiful-bees/?pid=7213&viewall=true

22. The Great Sunflower Project is an example of a citizen science project you can join by growing sunflowers and counting bees. https://www.greatsunflower.org/

23. Tracey, David. "Replanting the City Farming Movement in BC." *The Tyee*. August 20, 2009. http://thetyee.ca/News/2009/08/20/ReplantingFarming/

24. Victory Gardens 2007. http://www.futurefarmers.com/victorygardens/history.html

25. Smith, K. Annabelle. "A WWII Propaganda Campaign Popularized the Myth That Carrots Help You See in the Dark." *Smithsonian.com*. August 13, 2013. http://www.smithsonianmag.com/ist/?next=/arts-culture/a-wwii-propaganda-campaign-popularized-the-myth-that-carrots-help-you-see-in-the-dark-28812484/

26. Davies, Caroline. "Queen Turns Corner of Palace Backyard into an Allotment." *The Guardian*. June 14, 2009. http://www.theguardian.com/uk/2009/jun/14/queen-allotment-organic-gardening

27 Pearson, Gwen. "You're Worrying About the Wrong Bees." *WIRED*. April 29, 2015. http://www.wired.com/2015/04/youre-worrying-wrong-bees/

28 Christensen, Ken. "Could a Mushroom Save the Honeybee?" *PBS Newshour*. September 21, 2015. http://www.pbs.org/newshour/updates/mushroom-save-honeybee/

29 "Calgary Eyeopener: Urban Bees." *CBC*. June 9, 2014. http://www.cbc.ca/eyeopener/episode/2014/06/09/urban-bees/

30 Ibid.

31 Shepherd, Matthew. "Nests for Native Bees." The Xerces Society. http://www.xerces.org/wp-content/uploads/2008/11/nests_for_native_bees_fact_sheet_xerces_society.pdf

32 He was not interned, but had to give up his goose-hunting rifle. The local Mountie advised him to bury it in the backyard and dig it up when the war was over.

33 Helzer, Chris. "Bee Goggles." *The Prairie Ecologist*. October 1, 2013. http://prairieecologist.com/2013/10/01/bee-goggles/

34 Lohmiller, George and Becky. "Not-so-Common-Milkweed." *The Old Farmer's Almanac*. http://www.almanac.com/content/not-so-common-milkweed

35 Each province and state has its own beekeeping acts and regulations. All beekeepers, no matter what their methods, must adhere to these rules.

36 What we call "micronutrients" are really macronutrients from a bee's perspective.

37 Pearson, Gwen. "Royal Jelly Isn't What Makes a Queen Bee a Queen Bee." *WIRED*. September 2, 2015.

38 McNeil, M.E.A. "Marla Spivak: Getting Bees Back on Their Own Six Feet." *American Bee Journal* 150, no. 9 (2010): 856–860. http://www.meamcneil.com/Spivak%20I.pdf

39 Horn, Tammy. *Beeconomy: What Women and Bees Can Teach Us about Local Trade and the Global Market*. Lexington, KY: University Press of Kentucky, 2012. 202.

40 Ibid, 208.

41 Winston, Mark L. *Bee Time: Lessons from the Hive*. Cambridge, MA: Harvard University Press, 2014. 85–86.

42 Zink, Lindsay. "Concurrent effects of landscape context and managed pollinators on wild bee communities and canola (*Brassica napus* L.) pollen deposition." Thesis, Department of Biological Sciences, University of Calgary, January 2013. http://www.uoguelph.ca/canpolin/Publications/Zinkthesis2013_fixed.pdf

43 Pellett, Frank. "Anise Hyssop: Wonder Honey Plant." *American Bee Journal*, 1940.

44 Quinby, M. *Mysteries of Bee-keeping Explained. Containing the Result of Thirty-five Years' Experience, and Directions for Using the Movable Comb and Box-hive, Together with the Most Approved Methods of Propagating the Italian Bee*. New York: O. Judd, 1867. 80.

45 Pellett, Frank C. *American Honey Plants*. Hamilton, IL: American Bee Journal, 1920. 7.

46 "Farming with Sainfoin." LegumePlus. http://sainfoin.eu/farming-sainfoin

47 Dickinson, Emily. "To make a prairie (1755)." In Ralph W. Franklin, ed. *The Poems of Emily Dickinson*. Cambridge, MA: The Belknap Press of Harvard University Press, 1998.

48 Heinrich, Bernd. *Bumblebee Economics*. Cambridge. Harvard University Press, 1979. 101.

49 Davidson, John. Lecture "Honey Plants for Bees." February 1, 1920.

50 Ibid.

51 Turnbull, W.H. *One Hundred Years of Beekeeping in British Columbia, 1858–1958*. Vernon, BC: B.C. Honey Producers' Association, 1958. 20.

52 Sadly, during World War II, British farmers ripped out acres of hedgerows to grow food for the war effort.

53 Dogterom, Marguerite. "Want Bigger Blueberries? Make Your Bees Work Harder." Simon Fraser University Media & Public Relations. August 14, 1996. http://www.sfu.ca/archive-university-communications/pre2002archive/features/1996/August96/Blueberries.html

54 "Honey Bees Need Trees." Barcham: The Tree Specialists. http://www.barchampro.co.uk/honey-bees-need-trees

55 Beresford-Kroeger, Diana. *Arboretum America: A Philosophy of the Forest*. University of Michigan, 2003. 119.

56 Carlton, Marc. "Re-defining 'Native' Plants." In *The Pollinator Garden: About Plants, Pollinating Insects, and Gardening*. http://www.foxleas.com/re-defining-native-plants.asp

57 Siegel, Taggart, and Jon Betz. *Queen of the Sun: What Are the Bees Telling Us?* West Hoathly, UK: Clairview Books, 2011.

58 Mosby, Ian. "Food on the Home Front during the Second World War." *Wartime Canada*. http://wartimecanada.ca/essay/eating/food-home-front-during-second-world-war

Endnotes from photographs

1 "Audio Bee Booth, an amplified habitat installation for native, solitary bees and wasps," at Balls Falls Conservation Area, Ontario, Canada, 2013. The nest plank within the booth shows a leafcutter bee and potter wasp nest. Installation by Sarah Peebles, assisted by Rob Cruickshank, electronics; John Kuisma, woodworking; and Julie Kee, pyrography. It is one in a series of Audio Bee Booths and Cabinets, and is a project of Resonating Bodies. (http://resonatingbodies.wordpress.com/ and on Facebook)

INSPIRING AND USEFUL WEBSITES

Aganetha Dyck RCA · www.aganethadyck.ca

Beyond Pesticides (Check out the *BEE Protective Habitat Guide*.) · www.beyondpesticides.org

Border Free Bees (Cameron Cartiere and Nancy Holmes) · www.borderfreebees.com

BugGuide.net · www.bugguide.net

David Suzuki Foundation: Pollinators feed us. Let's protect them. (Check out downloadable pamphlets *Toronto Plant Guide for Attracting Pollinators* and *A Guide to Toronto's Pollinators*.) · www.davidsuzuki.org/issues/wildlife-habitat/projects/save-the-bees-and-butterflies

Environmental Youth Alliance · www.eya.ca

Feed the Bees (Earthwise Society and the Delta Chamber of Commerce) · www.feedthebees.org

The Great Sunflower Project · www.greatsunflower.org

Hartley Botanic: Make Your Own Bumble Bee Nest · www.hartley-botanic.co.uk/growing-tips/diy-bumble-bee-nest

Illinois Wildflowers (Dr. John Hilty) · www.illinoiswildflowers.info

Invasive Species Centre · www.invasivespeciescentre.ca

James Wong: A Scientist's Guide to Awesome Stuff to Grow · www.jameswong.co.uk

jasna guy: with/drawing and . . . · www.jasnaguy.wordpress.com

The Pollinator Garden: About plants, pollinating insects, and gardening (Marc Carlton) · www.foxleas.com

Pollinator Pathway (Sarah Bergmann) · www.pollinatorpathway.com

Pollinators of Native Plants by Heather Holm · www.pollinatorsnativeplants.com

Rebecca Chesney · www.rebeccachesney.com

Resonating Bodies (Sarah Peebles) · www.resonatingbodies.wordpress.com

Seeds of Diversity · www.seeds.ca

UC Berkeley Urban Bee Lab · www.helpabee.org

Xerces Society for Invertebrate Conservation · www.xerces.org

HELPFUL BLOGS

Bug Eric (Eric R. Eaton) · www.bugeric.blogspot.ca

Bug Squad: Happenings in the Insect World (Kathy Keatley Garvey) · www.ucanr.edu/blogs/bugsquad

Charismatic Minifauna (Gwen Pearson) · www.wired.com/category/science/science-blogs/charismaticminifauna

Honey Bee Suite: A Better Way to Bee (Rusty Burlew) · www.honeybeesuite.com

Ibycter (Sean McCann) · www.ibycter.com

The Prairie Ecologist (Chris Helzer) · www.prairieecologist.com

SEED COMPANIES

Baker Creek Heirloom Seeds (Mansfield, MO) · www.rareseeds.com

Prairie Moon Nursery (Winona, MN) ·
www.prairiemoon.com

Richters Herbs (Goodwood, ON) ·
www.richters.com

Salt Spring Seeds (Salt Spring Island, BC) ·
www.saltspringseeds.com

Seedhunt.com (Freedom, CA) · www.seedhunt.com

West Coast Seeds (Delta, BC) ·
www.westcoastseeds.com

Wildflower Farm (Coldwater, ON) ·
www.wildflowerfarm.com

ONLINE PUBLICATIONS

Agriculture and Agri-Food Canada. "Native Pollinators and Agriculture in Canada." http://publications.gc.ca/collections/collection_2014/aac-aafc/A59-12-2014-eng.pdf

Moisset, Beatriz, and Stephen Buchmann. "Bee Basics: An Introduction to Our Native Bees." http://www.fs.usda.gov/Internet/FSE_DOCUMENTS/stelprdb5306468.pdf

Ontario Horticultural Association. "Roadsides: A Guide to Creating a Pollinator Patch." http://www.gardenontario.org/subdomains/conservation/resources/guide.pdf

Schroeder, William. "Eco-Buffers: A High Density Agroforestry Design Using Native Species." http://www.fs.fed.us/rm/pubs/rmrs_p068/rmrs_p068_072_075.pdf

Sustainable Agriculture Research & Education. "Cover Cropping for Pollinators and Beneficial Insects." http://www.sare.org/Learning-Center/Bulletins/Cover-Cropping-for-Pollinators-and-Beneficial-Insects

USEFUL AND INSPIRATIONAL BOOKS

Beresford-Kroeger, Diana. *Arboretum America: A Philosophy of the Forest.* Ann Arbor: University of Michigan, 2003.

Buchmann, Stephen L., and Gary Paul Nabhan. *The Forgotten Pollinators.* Washington, DC: Island Press, 1996.

Droege, Sam, and Laurence Packer. *Bees: An Up-Close Look at Pollinators Around the World.* Minneapolis, MN: Voyageur Press, 2015.

Early, Jeremy. *My Side of the Fence: The Natural History of a Surrey Garden.* Surrey, BC: Jeremy Early, 2013.

Frankie, Gordon W., Robbin W. Thorp, Rollin E. Coville and Barbara Ertter. *California Bees and Blooms: A Guide for Gardeners and Naturalists.* Berkeley: Heyday, 2014.

Gardiner, Mary M. *Good Garden Bugs: Everything You Need to Know About Beneficial Predatory Insects.* Beverly, MA: Quarry Books, 2015.

Goldberger, Miriam. *Taming Wildflowers: Bringing the Beauty and Splendor of Nature's Blooms into Your Own Backyard.* Pittsburgh: St. Lynn's Press, 2014.

Grissell, Eric. *Bees, Wasps, and Ants: The Indispensable Role of Hymenoptera in Gardens.* Portland, OR: Timber Press, 2010.

Grissell, Eric. *Insects and Gardens: In Pursuit of a Garden Ecology.* Portland, OR: Timber Press, 2001.

Holm, Heather. *Pollinators of Native Plants: Attract, Observe and Identify Pollinators and Beneficial Insects with Native Plants.* Minnetonka, MN: Pollination Press, 2014.

Kirk, W.D.J., and F.N. Howes. *Plants for Bees: A Guide to the Plants that Benefit the Bees of the British Isles.* Cardiff, UK: International Bee Research Association, 2012.

LeBuhn, Gretchen. *Field Guide to the Common Bees of California.* Berkeley: University of California Press, 2013.

Packer, Laurence. *Keeping the Bees: Why All Bees Are at Risk and What We Can Do to Save Them.* Toronto: HarperCollins, 2010.

Walliser, Jessica. *Attracting Beneficial Bugs to Your Garden.* Portland, OR: Timber Press, 2014.

Williams, Paul H., Robbin W. Thorp, Leif L. Richardson and Sheila R. Colla. *Bumble Bees of North America: An Identification Guide.* Princeton, NJ: Princeton University Press, 2014.

Wilson, Joseph S., and Olivia Messinger Carril. *The Bees in Your Backyard: A Guide to North America's Bees.* Princeton, NJ: Princeton University Press, 2015.

The Xerces Society. *Attracting Native Pollinators: Protecting North America's Bees and Butterflies.* North Adams, MA: Storey Publishing, 2011.

INDEX

Opposite: Goldenrod is an essential bee garden plant,
providing food for a wide variety of bees and beneficial
insects in late summer and fall.

About the Author

*L*ori Weidenhammer is a Vancouver-based artist originally from Cactus Lake, SK. For nearly a decade she has been exploring the persona Madame Beespeaker, reviving the tradition of "telling the bees." She also appears as The Queen Bee at schools and community events. Lori is a member of the Second Site collective and Women Who Run with the Bees. Her collaborative media works with her partner Peter Courtemanche have been shown in Canada and abroad, and she has also worked in devised theatre in Canada and the UK. As an artist and educator, Weidenhammer works with students of all ages on growing and eating locally, planting for pollinators and community gardening. She is passionate about art that creates and strengthens community bonds and makes the world a better place for humans and bees.

We can do it! Let's work together to plant Victory Gardens for Bees. PETER COURTEMANCHE PHOTO